Memorable Quotations: Philosophers of Western Civilization

Compiled and Edited
by Carol A. Dingle

Writers Club Press
San Jose New York Lincoln Shanghai

Memorable Quotations: Philosophers of Western Civilization

Published by Writers Club Press
an imprint of iUniverse.com, Inc.

For information address:
iUniverse.com, Inc.
620 North 48th Street
Suite 201
Lincoln, NE 68504-3467
www.iuniverse.com

ISBN: 0-595-01183-7

Printed in the United States of America

Dedicated to my parents,

Leon and Lillian Dingle

Introduction

What is the meaning of existence? Philosophers throughout the ages have attempted to answer this question and understand the nature of being and thinking. With logical reasoning, they have searched for causes, basic truths, and principles about the universe, life, knowledge, and conduct.

In the tradition of Western civilization, philosophy is composed of different areas of speculation about various topics. Metaphysics examines the physical world and the individual's place in the universe. Logic probes the essence of knowledge. Epistemology seeks to ascertain the connection of human knowledge with the natural world. Ethics is concerned with morality. Aesthetics focuses on fundamentals of artistic expression.

Individual philosophers emphasize specific philosophical concepts. Idealism, or subjectivism, asserts that reality exists as thought rather than matter. Materialism deems matter and its motions as real. Dualism recognizes two realities, for example, mind and body, spirit and matter. Positivism contends that experience exists unto itself and is the basis for any action.

Historically, Western philosophy falls into three groups: Greek philosophy, primarily concerned about improving the city-state and the citizen; medieval philosophy, aligned with early Christian theology; and modern philosophy, which began in the Renaissance, leaning toward natural sciences. Various schools of thought have developed over time, including realism, romanticism, pessimism, transcendentalism, evolution, utilitarianism, positivism, pragmatism, and idealism. The words of insight and wisdom cited in this book cast light on the ultimate reality of life.

Nicola Abbagnano

(1901-90)
Italian philosopher

Reason itself is fallible, and this fallibility must find a place in our logic.

Theodor W. Adorno

(1903-69)
German philosopher

Art is permitted to survive only if it renounces the right to be different and integrates itself into the omnipotent realm of the profane.

Culture is only true when implicitly critical, and the mind which forgets this revenges itself in the critics it breeds. Criticism is an indispensable element of culture.

There are no more ideologies in the authentic sense of false consciousness, only advertisements for the world through its duplication and the provocative lie which does not seek belief but commands silence.

Alain [Émile-Auguste Chartier]

(1868-1951)
French philosopher

When we speak, in gestures or signs, we fashion a real object in the world; the gesture is seen, the words and the song are heard. The arts are simply a kind of writing, which, in one way or another, fixes words or gestures, and gives body to the invisible.

It is the human condition to question one god after another, one appearance after another, or better, one apparition after another, always pursuing the truth of the imagination, which is not the same as the truth of appearance.

Man himself is an enigma in motion; his questions never stay asked; whereas the mold, the footprint, and by natural extension, the statue itself, like the vaults, the arches, the temples with which man records his own passing, remain immobile and fix a moment of man's life, upon which one might endlessly meditate.

As opposed to the incoherent spectacle of the world, the real is what is expected, what is obtained and what is discovered by our own movement. It is what is sensed as being within our own power and always responsive to our action.

If religion is only human, and its form is man's form, it follows that everything in religion is true.

Henri-Frédéric Amiel

(1821-81)
Swiss philosopher, poet

Action is only coarsened thought—thought becomes concrete, obscure, and unconscious.

Our systems, perhaps, are nothing more than an unconscious apology for our faults—a gigantic scaffolding whose object is to hide from us our favorite sin.

Common sense is the measure of the possible; it is composed of experience and prevision; it is calculation applied to life.

An error is the more dangerous in proportion to the degree of truth which it contains.

To do easily what is difficult for others is the mark of talent. To do what is impossible for talent is the mark of genius.

It is by teaching that we teach ourselves, by relating that we *observe*, by affirming that we examine, by showing that we look, by writing that we think, by pumping that we draw water into the well.

The man who insists upon seeing with perfect clearness before he decides, never decides. Accept life, and you must accept regret.

Cleverness is serviceable for everything, sufficient for nothing.

Mutual respect implies discretion and reserve even in love itself; it means preserving as much liberty as possible to those whose life we share. We must distrust our instinct of intervention, for the desire to make one's own will prevail is often disguised under the mask of solicitude.

In every loving woman there is a priestess of the past—a pious guardian of some affection of which the object has disappeared.

Melancholy is at the bottom of everything, just as at the end of all rivers is the sea. Can it be otherwise in a world where nothing lasts, where all that we have loved or shall love must die? Is death, then, the secret of life? The gloom of an eternal mourning enwraps, more or less closely, every serious and thoughtful soul, as night enwraps the universe.

Mozart has the classic purity of light and the blue ocean; Beethoven the romantic grandeur which belongs to the storms of air and sea, and while the soul of Mozart seems to dwell on the ethereal peaks of Olympus, that of Beethoven climbs shuddering the storm-beaten sides of a Sinai. Blessed be they both! Each represents a moment of the ideal life, each does us good. Our love is due to both.

The obscure only exists that it may cease to exist. In it lies the opportunity of all victory and all progress. Whether it call itself fatality, death, night, or matter, it is the pedestal of life, of light, of liberty and the spirit. For it represents *resistance*—that is to say, the fulcrum of all activity, the occasion for its development and its triumph.

Without passion man is a mere latent force and possibility, like the flint which awaits the shock of the iron before it can give forth its spark.

The philosopher is like a man fasting in the midst of universal intoxication. He alone perceives the illusion of which all creatures are the willing playthings; he is less duped than his neighbor by his own nature. He judges more sanely, he sees things as they are. It is in this that his liberty consists—in the ability to see clearly and soberly, in the power of mental record.

Our true history is scarcely ever deciphered by others. The chief part of the drama is a monologue, or rather an intimate debate between God, our conscience, and ourselves. Tears, griefs, depressions, disappointments, irritations, good and evil thoughts, decisions, uncertainties, deliberations—all these belong to our secret, and are almost all incommunicable and intransmissible, even when we try to speak of them, and even when we write them down.

Self-interest is but the survival of the animal in us. Humanity only begins for man with self-surrender.

To know how to suggest is the great art of teaching. To attain it we must be able to guess what will interest; we must learn to read the childish soul as we might a piece of music. Then, by simply changing the key, we keep up the attraction and vary the song.

What we call little things are merely the causes of great things; they are the beginning, the embryo, and it is the point of departure which, generally speaking, decides the whole future of an existence. One single black speck may be the beginning of a gangrene, of a storm, of a revolution.

Thought is a kind of opium; it can intoxicate us, while still broad awake; it can make transparent the mountains and everything that exists.

Uncertainty is the refuge of hope.

Saint Thomas Aquinas

(1225-1274)
Italian scholastic philosopher

Law has the power to compel: indeed, the ability to enforce is a condition of the ability to command.

There is no leisure about politics.

For a war to be just, three conditions are necessary—public authority, just cause, right motive.

Hannah Arendt

(1906-75)
German-born U.S. political philosopher

No cause is left but the most ancient of all, the one, in fact, that from the beginning of our history has determined the very existence of politics, the cause of freedom versus tyranny.

No punishment has ever possessed enough power of deterrence to prevent the commission of crimes. On the contrary, whatever the punishment, once a specific crime has appeared for the first time, its reappearance is more likely than its initial emergence could ever have been.

Death not merely ends life, it also bestows upon it a silent completeness, snatched from the hazardous flux to which all things human are subject.

The earth is the very quintessence of the human condition.

Economic growth may one day turn out to be a curse rather than a good, and under no conditions can it either lead into freedom or constitute a proof for its existence.

The human condition is such that pain and effort are not just symptoms which can be removed without changing life itself; they are the modes in which life itself, together with the necessity to which it is bound, makes itself felt. For mortals, the "easy life of the gods" would be a lifeless life.

Love, by reason of its passion, destroys the in-between which relates us to and separates us from others. As long as its spell lasts, the only in-between which can insert itself between two lovers is the child, love's own product. The child, this in-between to which the lovers now are related and which they hold in common, is representative of the world in that it also separates them; it is an indication that they will insert a new world into the existing world. Through the child, it is as though the lovers return to the world from which their love had expelled them. But this new worldliness, the possible result and the only possibly happy ending of a love affair, is, in a sense, the end of love, which must either overcome the partners anew or be transformed into another mode of belonging together.

In contrast to revenge, which is the natural, automatic reaction to transgression and which, because of the irreversibility of the action process can be expected and even calculated, the act of forgiving can never be predicted; it is the only reaction that acts in an unexpected way and thus retains, though being a reaction, something of the original character of action.

It is quite gratifying to feel guilty if you haven't done anything wrong: how noble! Whereas it is rather hard and certainly depressing to admit guilt and to repent.

What makes it so plausible to assume that hypocrisy is the vice of vices is that integrity can indeed exist under the cover of all other vices except this one. Only crime and the criminal, it is true, confront us with the perplexity of radical evil; but only the hypocrite is really rotten to the core.

Immortality is what nature possesses without effort and without anybody's assistance, and immortality is what the mortals must therefore try to achieve if they want to live up to the world into which they were born, to live up to the things which surround them and to whose company they are admitted for a short while.

No civilization…would ever have been possible without a framework of stability, to provide the wherein for the flux of change. Foremost among the stabilizing factors, more enduring than customs, manners and traditions, are the legal systems that regulate our life in the world and our daily affairs with each other.

Love, by its very nature, is unworldly, and it is for this reason rather than its rarity that it is not only apolitical but antipolitical, perhaps the most powerful of all antipolitical human forces.

Total loyalty is possible only when fidelity is emptied of all concrete content, from which changes of mind might naturally arise.

Freedom from labor itself is not new; it once belonged among the most firmly established privileges of the few. In this instance, it seems as though scientific progress and technical developments had been only taken advantage of to achieve something about which all former ages dreamed but which none had been able to realize.

The point, as Marx saw it, is that dreams never come true.

Man cannot be free if he does not know that he is subject to necessity, because his freedom is always won in his never wholly successful attempts to liberate himself from necessity.

It is my contention that civil disobedients are nothing but the latest form of voluntary association, and that they are thus quite in tune with the oldest traditions of the country.

The new always happens against the overwhelming odds of statistical laws and their probability, which for all practical, everyday purposes amounts to certainty; the new therefore always appears in the guise of a miracle.

Opinions are formed in a process of open discussion and public debate, and where no opportunity for the forming of opinions exists, there may be moods—moods of the masses and moods of individuals, the latter no less fickle and unreliable than the former—but no opinion.

Poets...are the only people to whom love is not only a crucial, but an indispensable experience, which entitles them to mistake it for a universal one.

It is in the very nature of things human that every act that has once made its appearance and has been recorded in the history of mankind stays with mankind as a potentiality long after its actuality has become a thing of the past.

Predictions of the future are never anything but projections of present automatic processes and procedures, that is, of occurrences that are likely to come to pass if men do not act and if nothing unexpected happens; every action, for better or worse, and every accident necessarily destroys the whole pattern in whose frame the prediction moves and where it finds its evidence.

Promises are the uniquely human way of ordering the future, making it predictable and reliable to the extent that this is humanly possible.

There is all the difference in the world between the criminal's avoiding the public eye and the civil disobedient's taking the law into his own hands in open defiance. This distinction between an open violation of the law, performed in public, and a clandestine one is so glaringly obvious that it can be neglected only by prejudice or ill will.

The most radical revolutionary will become a conservative the day after the revolution.

Wherever the relevance of speech is at stake, matters become political by definition, for speech is what makes man a political being.

Only the mob and the elite can be attracted by the momentum of totalitarianism itself. The masses have to be won by propaganda.

Totalitarianism is never content to rule by external means, namely, through the state and a machinery of violence; thanks to its peculiar ideology and the role assigned to it in this apparatus of coercion, totalitarianism has discovered a means of dominating and terrorizing human beings from within.

The trouble with lying and deceiving is that their efficiency depends entirely upon a clear notion of the truth that the liar and deceiver wishes to hide. In this sense, truth, even if it does not prevail in public, possesses an ineradicable primacy over all falsehoods.

To be sure, nothing is more important to the integrity of the universities...than a rigorously enforced divorce from war-oriented research and all connected enterprises.

Power and violence are opposites; where the one rules absolutely, the other is absent. Violence appears where power is in jeopardy, but left to its own course it ends in power's disappearance.

Aristotle

(384-322 B.C.)
Greek philosopher

We praise a man who feels angry on the right grounds and against the right persons and also in the right manner at the right moment and for the right length of time.

Beauty depends on size as well as symmetry. No very small animal can be beautiful, for looking at it takes so small a portion of time that the impression of it will be confused. Nor can any very large one, for a whole view of it cannot be.

The beginning of reform is not so much to equalize property as to train the noble sort of natures not to desire more, and to prevent the lower from getting more.

What is a friend? A single soul dwelling in two bodies.

If happiness is activity in accordance with excellence, it is reasonable that it should be in accordance with the highest excellence.

All men by nature desire knowledge.

No great genius has ever existed without some touch of madness.

So it is naturally with the male and the female; the one is superior, the other inferior; the one governs, the other is governed; and the same rule must necessarily hold good with respect to all mankind.

The most perfect political community must be amongst those who are in the middle rank, and those states are best instituted wherein these are a larger and more respectable part, if possible, than both the other; or, if that cannot be, at least than either of them separate.

The moral virtues, then, are produced in us neither by nature nor against nature. Nature, indeed, prepares in us the ground for their reception, but their complete formation is the product of habit.

Man is by nature a political animal.

The generality of men are naturally apt to be swayed by fear rather than reverence, and to refrain from evil rather because of the punishment that it brings than because of its own foulness.

For as the interposition of a rivulet, however small, will occasion the line of the phalanx to fluctuate, so any trifling disagreement will be the cause of seditions; but they will not so soon flow from anything else as from the disagreement between virtue and vice, and next to that between poverty and riches.

Inferiors revolt in order that they may be equal, and equals that they may be superior. Such is the state of mind which creates revolutions.

Nor was civil society founded merely to preserve the lives of its members; but that they might live well: for otherwise a state might be composed of slaves, or the animal creation…nor is it an alliance mutually to defend each other from injuries, or for a commercial intercourse….But whosoever endeavors to establish wholesome laws in a state, attends to the virtues and vices of each individual who composes it; from whence it is evident, that the first care of him who would found a city, truly deserving that name, and not nominally so, must be to have his citizens virtuous.

What the statesman is most anxious to produce is a certain moral character in his fellow citizens, namely a disposition to virtue and the performance of virtuous actions.

Plato is dear to me, but dearer still is truth.

Wit is educated insolence.

Saint Augustine

(354-430)
Bishop of Hippo, theologian, philosopher

I have learnt to love you late, Beauty at once so ancient and so new!

The Devil invented gambling.

The desire for fame tempts even noble minds.

Marcus Aurelius

(121-80)
Roman emperor, philosopher

That which is not good for the beehive cannot be good for the bees.

A man should *be* upright, not be *kept* upright.

Let it be your constant method to look into the design of people's actions, and see what they would be at, as often as it is practicable; and to make this custom the more significant, practice it first upon yourself.

Sir Alfred J. Ayer

(1910-89)
British philosopher

To say that authority, whether secular or religious, supplies no ground for morality is not to deny the obvious fact that it supplies a sanction.

The ground for taking ignorance to be restrictive of freedom is that it causes people to make choices which they would not have made if they had seen what the realization of their choices involved.

While moral rules may be propounded by authority, the fact that these were so propounded would not validate them.

The traditional disputes of philosophers are, for the most part, as unwarranted as they are unfruitful.

There never comes a point where a theory can be said to be true. The most that one can claim for any theory is that it has shared the successes of all its rivals and that it has passed at least one test which they have failed.

Gaston Bachelard

(1884-1962)
French scientist, philosopher, literary theorist

Childhood lasts all through life. It returns to animate broad sections of adult life....Poets will help us to find this living childhood within us, this permanent, durable immobile world.

Even a minor event in the life of a child is an event of that child's world and thus a world event.

The subconscious is ceaselessly murmuring, and it is by listening to these murmurs that one hears the truth.

Man is a creation of desire, not a creation of need.

Reverie is not a mind vacuum. It is rather the gift of an hour which knows the plenitude of the soul.

There is no original truth, only original error.

If I were asked to name the chief benefit of the house, I should say: the house shelters day-dreaming, the house protects the dreamer, the house allows one to dream in peace.

Ideas are refined and multiplied in the commerce of minds. In their splendor, images effect a very simple communion of souls.

Ideas are invented only as correctives to the past. Through repeated rectifications of this kind one may hope to disengage an idea that is valid.

Man is an imagining being.

A special kind of beauty exists which is born in language, of language, and for language.

To live life well is to express life poorly; if one expresses life too well, one is living it no longer.

Literary imagination is an aesthetic object offered by a writer to a lover of books.

Two half philosophers will probably never a whole metaphysician make.

One must always maintain one's connection to the past and yet ceaselessly pull away from it. To remain in touch with the past requires a love of memory. To remain in touch with the past requires a constant imaginative effort.

Poetry is one of the destinies of speech....One would say that the poetic image, in its newness, opens a future to language.

To feel most beautifully alive means to be reading something beautiful, ready always to apprehend in the flow of language the sudden
flash of poetry.

The repose of sleep refreshes only the body. It rarely sets the soul at rest.
The repose of the night does not belong to us. It is not the possession of
our being. Sleep opens within us an inn for phantoms. In the morning
we must sweep out the shadows.

I am a dreamer of words, of written words. I think I am reading; a word
stops me. I leave the page. The syllables of the word begin to move
around. Stressed accents begin to invert. The word abandons its meaning
like an overload which is too heavy and prevents dreaming. Then words
take on other meanings as if they had the right to be young. And the
words wander away, looking in the nooks and crannies of vocabulary for
new company, bad company.

The words of the world want to make sentences.

A word is a bud attempting to become a twig. How can one not dream
while writing? It is the pen which dreams. The blank page gives the right
to dream.

Francis Bacon

(1561-1626)
English philosopher, essayist, statesman

There is as much difference between the counsel that a friend giveth,
and that a man giveth himself, as there is between the counsel of a
friend and of a flatterer. For there is no such flatterer as is a man's self.

I had rather believe all the Fables in the Legend, and the Talmud, and the Alcoran, than that this universal frame is without a Mind.

The pencil of the Holy Ghost hath laboured more in describing the afflictions of Job than the felicities of Solomon.

Prosperity is the blessing of the Old Testament; adversity is the blessing of the New.

Some books are to be tasted, others to be swallowed, and some few to be chewed and digested.

Nakedness is uncomely, as well in mind as body, and it addeth no small reverence to men's manners and actions if they be not altogether open....Therefore set it down: *That a habit of secrecy is both politic and moral.*

If a man will begin with certainties, he shall end in doubts, but if he will be content to begin with doubts, he shall end in certainties.

God's first creature, which was light.

I do not believe that any man fears to be dead, but only the stroke of death.

It is as natural to die as to be born; and to a little infant, perhaps, the one is as painful as the other.

The great advantages of simulation and dissimulation are three. First to lay asleep opposition and to surprise. For where a man's intentions are published, it is an alarum to call up all that are against them. The second is to reserve a man's self a fair retreat: for if a man engage himself, by a manifest declaration, he must go through, or take a fall. The third is, the better to discover the mind of another. For to him that opens himself, men will hardly show themselves adverse; but will fair let him go on, and turn their freedom of speech to freedom of thought.

They are ill discoverers that think there is no land when they see nothing but sea.

Discretion of speech is more than eloquence, and to speak agreeably to him with whom we deal is more than to speak in good words, or in good order.

He that hath wife and children hath given hostages to fortune; for they are impediments to great enterprises, either of virtue or mischief.

God Almighty first planted a garden. And indeed, it is the purest of human pleasures.

It is the true office of history to represent the events themselves, together with the counsels, and to leave the observations and conclusions thereupon to the liberty and faculty of every man's judgment.

Histories make men wise; poets witty; the mathematics subtle; natural philosophy deep; moral grave; logic and rhetoric able to contend.

Houses are built to live in, and not to look on: therefore let use be preferred before uniformity.

As the births of living creatures, at first, are ill-shapen: so are all *Innovations*, which are the births of time.

If we do not maintain Justice, Justice will not maintain us.

For a crowd is not company; and faces are but a gallery of pictures; and talk but a tinkling cymbal, where there is no love.

Certainly the best works, and of greatest merit for the public, have proceeded from the unmarried, or childless men.

Money is like muck, not good except it be spread.

Men of age object too much, consult too long, adventure too little, repent too soon, and seldom drive business home to the full period, but content themselves with a mediocrity of success.

We are much beholden to Machiavel and others, that write what men do, and not what they ought to do.

It is as hard and severe a thing to be a true politician as to be truly moral.

It is a strange desire, to seek power, and to lose liberty; or to seek power over others, and to lose power over a man's self.

Read not to contradict and confute; nor to believe and take for granted; nor to find talk and discourse; but to weigh and consider.

Revenge is a kind of wild justice, which the more a man's nature runs to, the more ought law to weed it out.

Whosoever is delighted in solitude is either a wild beast or a god.

There is nothing makes a man suspect much, more than to know little.

Reading maketh a full man; conference a ready man; and writing an exact man.

It is a pleasure to stand upon the shore, and to see ships tost upon the sea: a pleasure to stand in the window of a castle, and to see a battle and the adventures thereof below: but no pleasure is comparable to standing upon the vantage ground of truth…and to see the errors, and wanderings, and mists, and tempests, in the vale below.

For it is not possible to join serpentine wisdom with columbine innocency, except men know exactly all the conditions of the serpent: his baseness and going upon his belly, his volubility and lubricity, his envy and sting, and the rest; that is, all forms and natures of evil: for without this, virtue lieth open and unfenced.

Wives are young men's mistresses, companions for middle age, and old men's nurses.

Roger Bacon

(c.1214-1294?)
English philosopher, scientist

Argument is conclusive…but…it does not remove doubt, so that the mind may rest in the sure knowledge of the truth, unless it finds it by the method of experiment.…For if any man who never saw fire proved by satisfactory arguments that fire burns…his hearer's mind would never be satisfied, nor would he avoid the fire until he put his hand in it…that he might learn by experiment what argument taught.

All science requires mathematics. The knowledge of mathematical things is almost innate in us.…This is the easiest of sciences, a fact which is obvious in that no one's brain rejects it; for laymen and people who are utterly illiterate know how to count and reckon.

Simone de Beauvoir

(1908-86)
French novelist, essayist, philosopher

To *make* oneself an object, to *make* oneself passive, is a very different thing from *being* a passive object.

In order for the artist to have a world to express he must first be situated in this world, oppressed or oppressing, resigned or rebellious, a man among men.

The curse which lies upon marriage is that too often the individuals are joined in their weakness rather than in their strength—each asking from the other instead of finding pleasure in giving. It is even more deceptive to dream of gaining through the child a plenitude, a warmth, a value, which one is unable to create for oneself; the child brings joy only to the woman who is capable of disinterestedly desiring the happiness of another, to one who without being wrapped up in self seeks to transcend her own existence.

Since it is the Other within us who is old, it is natural that the revelation of our age should come to us from outside—from others. We do not accept it willingly.

It is old age, rather than death, that is to be contrasted with life. Old age is life's parody, whereas death transforms life into a destiny: in a way it preserves it by giving it the absolute dimension. Death does away with time.

Sex pleasure in woman is a kind of magic spell; it demands complete abandon; if words or movements oppose the magic of caresses, the spell is broken.

One is not born, but rather becomes, a woman.

Walter Benjamin

(1892-1940)
German critic, philosopher

Separation penetrates the disappearing person like a pigment and steeps him in gentle radiance.

Counsel woven into the fabric of real life is wisdom.

The greater the decrease in the social significance of an art form, the sharper the distinction between criticism and enjoyment by the public. The conventional is uncritically enjoyed, and the truly new is criticized with aversion.

Reminiscences, even extensive ones, do not always amount to an auto-biography....For autobiography has to do with time, with sequence and what makes up the continuous flow of life. Here, I am talking of a space, of moments and discontinuities. For even if months and years appear here, it is in the form they have in the moment of recollection. This strange form—it may be called fleeting or eternal—is in neither case the stuff that life is made of.

To be happy is to be able to become aware of oneself without fright.

The idea that happiness could have a share in beauty would be too much of a good thing.

All religions have honored the beggar. For he proves that in a matter at the same time as prosaic and holy, banal and regenerative as the giving of alms, intellect and morality, consistency and principles are miserably inadequate.

Of all the ways of acquiring books, writing them oneself is regarded as the most praiseworthy method....Writers are really people who write books not because they are poor, but because they are dissatisfied with the books which they could buy but do not like.

Boredom is the dream bird that hatches the egg of experience. A rustling in the leaves drives him away.

The book borrower of real stature whom we envisage here proves himself to be an inveterate collector of books not so much by the fervor with which he guards his borrowed treasures and by the deaf ear which he turns to all reminders from the everyday world of legality as by his failure to read these books.

The killing of a criminal can be moral—but never its legitimation.

Not to find one's way in a city may well be uninteresting and banal. It requires ignorance—nothing more. But to lose oneself in a city—as one loses oneself in a forest—that calls for a quite different schooling. Then, signboard and street names, passers-by, roofs, kiosks, or bars must speak to the wanderer like a cracking twig under his feet in the forest.

Experience has taught me that the shallowest of communist platitudes contains more of a hierarchy of meaning than contemporary bourgeois profundity.

These are days when no one should rely unduly on his "competence." Strength lies in improvisation. All the decisive blows are struck left-handed.

The art of the critic in a nutshell: to coin slogans without betraying ideas. The slogans of an inadequate criticism peddle ideas to fashion.

Each morning the day lies like a fresh shirt on our bed; this incomparably fine, incomparably tightly woven tissue of pure prediction fits us perfectly. The happiness of the next twenty-four hours depends on our ability, on waking, to pick it up.

Books and harlots have their quarrels in public.

The destructive character lives from the feeling, not that life is worth living, but that suicide is not worth the trouble.

Only he who can view his own past as an abortion sprung from compulsion and need can use it to full advantage in the present. For what one has lived is at best comparable to a beautiful statue which has had all its limbs knocked off in transit, and now yields nothing but the precious block out of which the image of one's future must be hewn.

The construction of life is at present in the power of facts far more than convictions.

Taking food alone tends to make one hard and coarse. Those accustomed to it must lead a Spartan life if they are not to go downhill. Hermits have observed, if for only this reason, a frugal diet. For it is only in company that eating is done justice; food must be divided and distributed if it is to be well received.

Gifts must affect the receiver to the point of shock.

All human knowledge takes the form of interpretation.

We have long forgotten the ritual by which the house of our life was erected. But when it is under assault and enemy bombs are already taking their toll, what enervated, perverse antiquities do they not lay bare in the foundations.

Genuine polemics approach a book as lovingly as a cannibal spices a baby.

The adjustment of reality to the masses and of the masses to reality is a process of unlimited scope, as much for thinking as for perception.

Like ultraviolet rays memory shows to each man in the book of life a script that invisibly and prophetically glosses the text.

Memory is not an instrument for exploring the past but its theatre. It is the medium of past experience, as the ground is the medium in which dead cities lie interred.

Opinions are to the vast apparatus of social existence what oil is to machines: one does not go up to a turbine and pour machine oil over it; one applies a little to hidden spindles and joints that one has to know.

The true picture of the past flits by. The past can be seized only as an image which flashes up at the instant when it can be recognized and is never seen again.

The camera introduces us to unconscious optics as does psychoanalysis to unconscious impulses.

He who asks fortune-tellers the future unwittingly forfeits an inner intimation of coming events that is a thousand times more exact than anything they may say. He is impelled by inertia, rather than curiosity, and nothing is more unlike the submissive apathy with which he hears his fate revealed than the alert dexterity with which the man of courage lays hands on the future.

It is precisely the purpose of the public opinion generated by the press to make the public incapable of judging, to insinuate into it the attitude of someone irresponsible, uninformed.

Bourgeois existence is the regime of private affairs…and the family is the rotten, dismal edifice in whose closets and crannies the most ignominious instincts are deposited. Mundane life proclaims the total subjugation of eroticism to privacy.

Living substance conquers the frenzy of destruction only in the ecstasy of procreation.

Opinions are a private matter. The public has an interest only in judgments.

Quotations in my work are like wayside robbers who leap out armed and relieve the stroller of his conviction.

The power of a text is different when it is read from when it is copied out....Only the copied text thus commands the soul of him who is occupied with it, whereas the mere reader never discovers the new aspects of his inner self that are opened by the text, that road cut through the interior jungle forever closing behind it: because the reader follows the movement of his mind in the free flight of day-dreaming, whereas the copier submits it to command.

The only way of knowing a person is to love them without hope.

He who seeks to approach his own buried past must conduct himself like a man digging....He must not be afraid to return again and again to the same matter; to scatter it as one scatters earth, to turn it over as one turns over soil. For the matter itself is only a deposit, a stratum, which yields only to the most meticulous examination what constitutes the real treasure hidden within the earth: the images, severed from all earlier associations, that stand—like precious fragments or torsos in a collector's gallery—in the prosaic rooms of our later understanding.

The art of storytelling is reaching its end because the epic side of truth, wisdom, is dying out.

Death is the sanction of everything the storyteller can tell. He has borrowed his authority from death.

Every passion borders on the chaotic, but the collector's passion borders on the chaos of memories.

All disgust is originally disgust at touching.

Any translation which intends to perform a transmitting function cannot transmit anything but information—hence, something inessential. This is the hallmark of bad translations.

Nothing is poorer than a truth expressed as it was thought. Committed to writing in such cases, it is not even a bad photograph....Truth wants to be startled abruptly, at one stroke, from her self-immersion, whether by uproar, music or cries for help.

Work on good prose has three steps: a musical stage when it is composed, an architectonic one when it is built, and a textile one when it is woven.

Jeremy Bentham

(1748-1832)
English philosopher, jurist, political theorist

The principle of asceticism never was, nor ever can be, consistently pursued by any living creature. Let but one tenth part of the inhabitants of the earth pursue it consistently, and in a day's time they will have turned it into a Hell.

The said truth is that it is the greatest happiness of the greatest number that is the measure of right and wrong.

Nicolai A. Berdyaev

(1874-1948)
Russian Christian philosopher

We find the most terrible form of atheism, not in the militant and passionate struggle against the idea of God himself, but in the practical atheism of everyday living, in indifference and torpor. We often encounter these forms of atheism among those who are formally Christians.

The bourgeois takes economic power very seriously, and often worships it quite unselfishly.

In sex we have the source of man's true connection with the cosmos and of his servile dependence. The categories of sex, male and female, are cosmic categories, not merely anthropological categories.

Henri Bergson

(1859-1941)
French philosopher

An absolute can only be given in an *intuition,* while all the rest has to do with *analysis.* We call intuition here the *sympathy* by which one is transported into the interior of an object in order to coincide with what there is unique and consequently inexpressible in it. Analysis, on the contrary, is the operation which reduces the object to elements already known.

To perceive means to immobilize...we seize, in the act of perception, something which outruns perception itself.

Spirit borrows from matter the perceptions on which it feeds and restores them to matter in the form of movements which it has stamped with its own freedom.

F. H. Bradley

(1846-1924)
English philosopher

The secret of happiness is to admire without desiring. And that is not happiness.

Our live experiences, fixed in aphorisms, stiffen into cold epigrams. Our heart's blood, as we write it, turns to mere dull ink.

The man who has ceased to fear has ceased to care.

We say that a girl with her doll anticipates the mother. It is more true, perhaps, that most mothers are still but children with playthings.

Metaphysics is the finding of bad reasons for what we believe upon instinct; but to find these reasons is no less an instinct.

There are those who so dislike the nude that they find something indecent in the naked truth.

The world is the best of all possible worlds, and everything in it is a necessary evil.

True penitence condemns to silence. What a man is ready to recall he would be willing to repeat.

There are persons who, when they cease to shock us, cease to interest us.

The one self-knowledge worth having is to know one's own mind.

It is by a wise economy of nature that those who suffer without change, and whom no one can help, become uninteresting. Yet so it may happen that those who need sympathy the most often attract it the least.

The force of the blow depends on the resistance. It is sometimes better not to struggle against temptation. Either fly or yield at once.

Eclecticism. Every truth is so true that any truth must be false.

The deadliest foe to virtue would be complete self-knowledge.

Giordano Bruno

(1548-1600)
Italian philosopher

We delight in one knowable thing, which comprehends all that is knowable; in one apprehensible, which draws together all that can be apprehended; in a single being that includes all, above all in the one which is itself the all.

It may be you fear more to deliver judgment upon me than I fear judgment.

The beginning, middle, and end of the birth, growth, and perfection of whatever we behold is from contraries, by contraries, and to contraries; and whatever contrarity is, there is action and reaction, there is motion, diversity, multitude, and order, there are degrees, succession and vicissitude.

The universe is then one, infinite, immobile....It is not capable of comprehension and therefore is endless and limitless, and to that extent infinite and indeterminable, and consequently immobile.

Edmund Burke

(1729-97)
Irish philosopher, statesman

When bad men combine, the good must associate; else they will fall, one by one, an unpitied sacrifice in a contemptible struggle.

Ambition can creep as well as soar.

We must not always judge of the generality of the opinion by the noise of the acclamation.

Nobility is a graceful ornament to the civil order. It is the Corinthian capital of polished society.

In the weakness of one kind of authority, and in the fluctuation of all, the officers of an army will remain for some time mutinous and full of faction, until some popular general, who understands the art of conciliating the soldiery, and who possesses the true spirit of command, shall draw the eyes of all men upon himself. Armies will obey him on his personal account. There is no other way of securing military obedience in this state of things.

Beauty in distress is much the most affecting beauty.

It is the interest of the commercial world that wealth should be found everywhere.

Men are qualified for civil liberty in exact proportion to their disposition to put moral chains upon their own appetites; in proportion as their love of justice is above their rapacity; in proportion as their soundness and sobriety of understanding is above their vanity and presumption; in proportion as they are more disposed to listen to the counsels of the wise and good, in preference to the flattery of knaves.

It is a general popular error to suppose the loudest complainers for the public to be the most anxious for its welfare.

All government—indeed every human benefit and enjoyment, every virtue and every prudent act—is founded on compromise and barter.

Among a people generally corrupt, liberty cannot long exist.

Under the pressure of the cares and sorrows of our mortal condition, men have at all times, and in all countries, called in some physical aid to their moral consolations—wine, beer, opium, brandy, or tobacco.

Mere parsimony is not economy....Expense, and great expense, may be an essential part in true economy.

When the leaders choose to make themselves bidders at an auction of popularity, their talents, in the construction of the state, will be of no service. They will become flatterers instead of legislators; the instruments, not the guides, of the people.

It is the nature of all greatness not to be exact.

No passion so effectually robs the mind of all its powers of acting and reasoning as fear.

The objects of a financier are, then, to secure an ample revenue; to impose it with judgment and equality; to employ it economically; and, when necessity obliges him to make use of credit, to secure its foundations in that instance, and for ever, by the clearness and candour of his proceedings, the exactness of his calculations, and the solidity of his funds.

Flattery corrupts both the receiver and the giver.

The use of force alone is but *temporary*. It may subdue for a moment; but it does not remove the necessity of subduing again: and a nation is not governed, which is perpetually to be conquered.

Nothing turns out to be so oppressive and unjust as a feeble government.

The great must submit to the dominion of prudence and of virtue, or none will long submit to the dominion of the great.

To innovate is not to reform.

A good parson once said that where mystery begins religion ends. Cannot I say, as truly at least, of human laws, that where mystery begins justice ends?

Laws, like houses, lean on one another.

In effect, to follow, not to force the public inclination; to give a direction, a form, a technical dress, and a specific sanction, to the general sense of the community, is the true end of legislature.

There is but one law for all, namely that law which governs all law, the law of our Creator, the law of humanity, justice, equity—the law of nature and of nations.

The true danger is when liberty is nibbled away, for expedience, and by parts.

The effect of liberty to individuals is that they may do what they please: we ought to see what it will please them to do, before we risk congratulations.

In doing good, we are generally cold, and languid, and sluggish; and of all things afraid of being too much in the right. But the works of malice and injustice are quite in another style. They are finished with a bold, masterly hand; touched as they are with the spirit of those vehement passions that call forth all our energies, whenever we oppress and persecute.

Manners are of more importance than laws....Manners are what vex or soothe, corrupt or purify, exalt or debase, barbarize or refine us, by a constant, steady, uniform, insensible operation, like that of the air we breathe in.

The tyranny of a multitude is a multiplied tyranny.

A nation is not conquered which is perpetually to be conquered.

He that wrestles with us strengthens our nerves, and sharpens our skill. Our antagonist is our helper. This amicable conflict with difficulty helps us to an intimate acquaintance with our object, and compels us to consider it in all its relations. It will not suffer us to be superficial.

I have never yet seen any plan which has not been mended by the observations of those who were much inferior in understanding to the person who took the lead in the business.

Those who have been once intoxicated with power, and have derived any kind of emolument from it, even though but for one year, never can willingly abandon it. They may be distressed in the midst of all their power; but they will never look to anything but power for their relief.

Nothing is so fatal to religion as indifference which is, at least, half infidelity.

Whilst shame keeps its watch, virtue is not wholly extinguished in the heart; nor will moderation be utterly exiled from the minds of tyrants.

Society is indeed a contract....It is a partnership in all science; a partnership in all art; a partnership in every virtue, and in all perfection. As the ends of such a partnership cannot be obtained in many generations, it becomes a partnership not only between those who are living, but between those who are living, those who are dead, and those who are to be born.

A state without the means of some change is without the means of its conservation.

A disposition to preserve, and an ability to improve, taken together, would be my standard of a statesman.

Superstition is the religion of feeble minds.

To tax and to please, no more than to love and to be wise, is not given to men.

People will not look forward to posterity, who never look backward to their ancestors.

Kings will be tyrants from policy, when subjects are rebels from principle.

It is, generally, in the season of prosperity that men discover their real temper, principles, and designs.

To drive men from independence to live on alms, is itself great cruelty.

The arrogance of age must submit to be taught by youth.

Albert Camus

(1913-60)
French-Algerian philosopher

At any street corner the feeling of absurdity can strike any man in the face.

It is impossible to give a clear account of the world, but art can teach us to reproduce it—just as the world reproduces itself in the course of its eternal gyrations. The primordial sea indefatigably repeats the same words and casts up the same astonished beings on the same seashore.

At the heart of all beauty lies something inhuman, and these hills, the softness of the sky, the outline of these trees at this very minute lose the illusory meaning with which we had clothed them, henceforth more remote than a lost paradise…that denseness and that strangeness of the world is absurd.

What will be left of the power of example if it is proved that capital punishment has another power, and a very real one, which degrades men to the point of shame, madness, and murder?

We are not certain, we are never certain. If we were we could reach some conclusions, and we could, at last, make others take us seriously.

You know what charm is: a way of getting the answer yes without having asked any clear question.

As a remedy to life in society I would suggest the big city. Nowadays, it is the only desert within our means.

A sub-clerk in the post-office is the equal of a conqueror if consciousness is common to them.

For centuries the death penalty, often accompanied by barbarous refinements, has been trying to hold crime in check; yet crime persists. Why? Because the instincts that are warring in man are not, as the law claims, constant forces in a state of equilibrium.

Culture: the cry of men in face of their destiny.

Without culture, and the relative freedom it implies, society, even when perfect, is but a jungle. This is why any authentic creation is a gift to the future.

Men are never really willing to die except for the sake of freedom: therefore they do not believe in dying completely.

There will be no lasting peace either in the heart of individuals or in social customs until death is outlawed.

True debauchery is liberating because it creates no obligations. In it you possess only yourself; hence it remains the favorite pastime of the great lovers of their own person.

To those who despair of everything reason cannot provide a faith, but only passion, and in this case it must be the same passion that lay at the root of the despair, namely humiliation and hatred.

Lucifer also has died with God, and from his ashes has arisen a spiteful demon who does not even understand the object of his venture.

You will never be happy if you continue to search for what happiness consists of. You will never live if you are looking for the meaning of life.

Man is the only creature who refuses to be what he is.

Europe has lived on its contradictions, flourished on its differences, and, constantly transcending itself thereby, has created a civilization on which the whole world depends even when rejecting it. This is why I do not believe in a Europe unified under the weight of an ideology or of a technocracy that overlooked these differences.

In order to exist just once in the world, it is necessary never again to exist.

We all carry within us our places of exile, our crimes, and our ravages. But our task is not to unleash them on the world; it is to fight them in ourselves and in others.

A novel is never anything but a philosophy put into images.

Absolute virtue is impossible and the republic of forgiveness leads, with implacable logic, to the republic of the guillotine.

The only conception of freedom I can have is that of the prisoner or the individual in the midst of the State. The only one I know is freedom of thought and action.

Without freedom, no art; art lives only on the restraints it imposes on itself, and dies of all others. But without freedom, no socialism either, except the socialism of the gallows.

The French Revolution gave birth to no artists but only to a great journalist, Desmoulins, and to an under-the-counter writer, Sade. The only poet of the times was the guillotine.

The gods had condemned Sisyphus to ceaselessly rolling a rock to the top of a mountain, whence the stone would fall back of its own weight. They had thought with some reason that there is no more dreadful punishment than futile and hopeless labor.

Real generosity towards the future lies in giving all to the present.

It is normal to give away a little of one's life in order not to lose it all.

If man is reduced to being nothing but a character in history, he has no other choice but to subside into the sound and fury of a completely irrational history or to endow history with the form of human reason.

History, as an entirety, could only exist in the eyes of an observer outside it and outside the world. History, only exists, in the final analysis, for God.

As for Hitler, his professed religion unhesitatingly juxtaposed the God-Providence and Valhalla. Actually his god was an argument at a political meeting and a manner of reaching an impressive climax at the end of speeches.

Instead of killing and dying in order to produce the being that we are not, we have to live and let live in order to create what we are.

Methods of thought which claim to give the lead to our world in the name of revolution have become, in reality, ideologies of consent and not of rebellion.

More and more, when faced with the world of men, the only reaction is one of individualism. Man alone is an end unto himself. Everything one tries to do for the common good ends in failure.

To live is to hurt others, and through others, to hurt oneself. Cruel earth! How can we manage not to touch anything? To find what ultimate exile?

Children will still die unjustly even in a perfect society. Even by his greatest effort, man can only propose to diminish, arithmetically, the sufferings of the world.

Absolute justice is achieved by the suppression of all contradiction: therefore it destroys freedom.

Modern conquerors can kill, but do not seem to be able to create. Artists know how to create but cannot really kill. Murderers are only very exceptionally found among artists.

Man wants to live, but it is useless to hope that this desire will dictate all his actions.

Accept life, take it as it is? Stupid. The means of doing otherwise? Far from our having to take it, it is life that possesses us and on occasion shuts our mouths.

We get into the habit of living before acquiring the habit of thinking. In that race which daily hastens us towards death, the body maintains its irreparable lead.

The desire for possession is insatiable, to such a point that it can survive even love itself. To love, therefore, is to sterilize the person one loves.

Martyrs, *cher ami*, must choose between being forgotten, mocked or made use of. As for being understood—never!

Just as all thought, and primarily that of non-signification, signifies something, so there is no art that has no signification.

The modern mind is in complete disarray. Knowledge has stretched itself to the point where neither the world nor our intelligence can find any foot-hold. It is a fact that we are suffering from nihilism.

After all, every murderer when he kills runs the risk of the most dreadful of deaths, whereas those who kill him risk nothing except promotion.

Truly fertile Music, the only kind that will move us, that we shall truly appreciate, will be a Music conducive to Dream, which banishes all reason and analysis. One must not wish first to understand and then to feel. Art does not tolerate Reason.

It is a well-known fact that we always recognize our homeland when we are about to lose it.

If only nature is real and if, in nature, only desire and destruction are legitimate, then, in that all humanity does not suffice to assuage the thirst for blood, the path of destruction must lead to universal annihilation.

We come into the world laden with the weight of an infinite necessity.

Manhattan. Sometimes from beyond the skyscrapers, across the hundreds of thousands of high walls, the cry of a tugboat finds you in your insomnia in the middle of the night, and you remember that this desert of iron and cement is an island.

Nihilism is not only despair and negation, but above all the desire to despair and to negate.

The world is never quiet, even its silence eternally resounds with the same notes, in vibrations which escape our ears. As for those that we perceive, they carry sounds to us, occasionally a chord, never a melody.

Only a philosophy of eternity, in the world today, could justify non-violence.

Those who weep for the happy periods which they encounter in history acknowledge what they want; not the alleviation but the silencing of misery.

If Christianity is pessimistic as to man, it is optimistic as to human destiny. Well, I can say that, pessimistic as to human destiny, I am optimistic as to man.

The Poor Man whom everyone speaks of, the Poor Man whom everyone pities, one of the repulsive Poor from whom "charitable" souls keep their distance, he has still said nothing. Or, rather, he has spoken through the voice of Victor Hugo, Zola, Richepin. At least, they said so. And these shameful impostures fed their authors. Cruel irony, the Poor Man tormented with hunger feeds those who plead his case.

From Paul to Stalin, the popes who have chosen Caesar have prepared the way for Caesars who quickly learn to despise popes.

The principles which men give to themselves end by overwhelming their noblest intentions.

To abandon oneself to principles is really to die—and to die for an impossible love which is the contrary of love.

The society based on production is only productive, not creative.

Retaliation is related to nature and instinct, not to law. Law, by definition, cannot obey the same rules as nature.

Realism should only be the means of expression of religious genius...or, at the other extreme, the artistic expressions of monkeys which are quite satisfied with mere imitation. In fact, art is never realistic though sometimes it is tempted to be. To be really realistic a description would have to be endless.

Every act of rebellion expresses a nostalgia for innocence and an appeal to the essence of being.

The rebel can never find peace. He knows what is good and, despite himself, does evil. The value which supports him is never given to him once and for all—he must fight to uphold it, unceasingly.

Human relationships always help us to carry on because they always presuppose further developments, a future—and also because we live as if our only task was precisely to have relationships with other people.

More and more, revolution has found itself delivered into the hands of its bureaucrats and doctrinaires on the one hand, and to the enfeebled and bewildered masses on the other.

Revolution, in order to be creative, cannot do without either a moral or metaphysical rule to balance the insanity of history.

Every revolutionary ends by becoming either an oppressor or a heretic.

To know oneself, one should assert oneself. Psychology is action, not thinking about oneself. We continue to shape our personality all our life. If we knew ourselves perfectly, we should die.

Ah, *mon cher*, for anyone who is alone, without God and without a master, the weight of days is dreadful.

In default of inexhaustible happiness, eternal suffering would at least give us a destiny. But we do not even have that consolation, and our worst agonies come to an end one day.

There is but one truly serious philosophical problem and that is suicide. Judging whether life is or is not worth living amounts to answering the fundamental question of philosophy. All the rest—whether or not the world has three dimensions, whether the mind has nine or twelve categories—comes afterwards. These are games; one must first answer.

Men are never convinced of your reasons, of your sincerity, of the seriousness of your sufferings, except by your death. So long as you are alive, your case is doubtful; you have a right only to your skepticism.

The society of merchants can be defined as a society in which things disappear in favor of signs. When a ruling class measures its fortunes, not by the acre of land or the ingot of gold, but by the number of figures corresponding ideally to a certain number of exchange operations, it thereby condemns itself to setting a certain kind of humbug at the center of its experience and its universe. A society founded on signs is, in its essence, an artificial society in which man's carnal truth is handled as something artificial.

To insure the adoration of a theorem for any length of time, faith is not enough, a police force is needed as well.

One leader, one people, signifies one master and millions of slaves.

Virtue cannot separate itself from reality without becoming a principle of evil.

We used to wonder where war lived, what it was that made it so vile. And now we realize that we know where it lives, that it is inside ourselves.

Elias Canetti

(1905-94)
Austrian novelist, philosopher

When you write down your life, every page should contain something no one has ever heard about.

Adults find pleasure in deceiving a child. They consider it necessary, but they also enjoy it. The children very quickly figure it out and then practice deception themselves.

The fear of burglars is not only the fear of being robbed, but also the fear of a sudden and unexpected clutch out of the darkness.

He who is obsessed by death is made guilty by it.

Every decision is liberating, even if it leads to disaster. Otherwise, why do so many people walk upright and with open eyes into their misfortune?

Whether or not God is dead: it is impossible to keep silent about him who was there for so long.

As if one could know the good a person is capable of, when one doesn't know the bad he might do.

There is no doubt: the study of man is just beginning, at the same time that his end is in sight.

It doesn't matter how new an idea *is*: what matters is how new it *becomes*.

Justice begins with the recognition of the necessity of sharing. The oldest law is that which regulates it, and this is still the most important law today and, as such, has remained the basic concern of all movements which have at heart the community of human activities and of human existence in general.

There is no such thing as an ugly language. Today I hear every language as if it were the only one, and when I hear of one that is dying, it overwhelms me as though it were the death of the earth.

Someone who always has to lie discovers that every one of his lies is true.

One should not confuse the craving for life with endorsement of it.

The profoundest thoughts of the philosophers have something tricklike about them. A lot disappears in order for something to suddenly appear in the palm of the hand.

The self-explorer, whether he wants to or not, becomes the explorer of everything else. He learns to see himself, but suddenly, provided he was honest, all the rest appears, and it is as rich as he was, and, as a final crowning, richer.

Success is the space one occupies in the newspaper. Success is one day's insolence.

There is nothing that man fears more than the touch of the unknown. He wants to *see* what is reaching towards him, and to be able to recognize or at least classify it. Man always tends to avoid physical contact with anything strange.

The process of writing has something infinite about it. Even though it is interrupted each night, it is one single notation.

Pierre Charron

(1541-1603)
French philosopher

The most excellent and divine counsel, the best and most profitable advertisement of all others, but the least practiced, is to study and learn how to know ourselves. This is the foundation of wisdom and the highway to whatever is good....God, Nature, the wise, the world, preach man, exhort him both by word and deed to the study of himself.

The true science and study of man is man himself.

Nikolai Gavrilovich Chernyshevsky

(1828-89)
Russian writer, philosopher

History is fond of her grandchildren, for it offers them the marrow of the bones, which the previous generation had hurt its hands in breaking.

Cicero

(106 B.C.-43 B.C.)
Roman orator, philosopher

Since an intelligence common to us all makes things known to us and formulates them in our minds, honorable actions are ascribed by us to virtue, and dishonorable actions to vice; and only a madman would conclude that these judgments are matters of opinion, and not fixed by nature.

Justice consists in doing no injury to men; decency in giving them no offense.

The good of the people is the greatest law.

No one is so old as to think he cannot live one more year.

I prefer tongue-tied knowledge to ignorant loquacity.

The sinews of war, a limitless supply of money.

E. M. Cioran

(1911-95)
Rumanian-born French philosopher

Does our ferocity not derive from the fact that our instincts are all too interested in other people? If we attended more to ourselves and became the center, the object of our murderous inclinations, the sum of our intolerances would diminish.

Those who believe in *their* truth—the only ones whose imprint is retained by the memory of men—leave the earth behind them strewn with corpses. Religions number in their ledgers more murders than the bloodiest tyrannies account for, and those whom humanity has called divine far surpass the most conscientious murderers in their thirst for slaughter.

Consciousness is much more than the thorn, it is the *dagger* in the flesh.

A sudden silence in the middle of a conversation suddenly brings us back to essentials: it reveals how dearly we must pay for the invention of speech.

Show me one thing here on earth which has begun well and not ended badly. The proudest palpitations are engulfed in a sewer, where they cease throbbing, as though having reached their natural term: this downfall constitutes the heart's drama and the negative meaning of history.

A decadent civilization compromises with its disease, cherishes the virus infecting it, loses its self-respect.

The source of our actions resides in an unconscious propensity to regard ourselves as the center, the cause, and the conclusion of time. Our reflexes and our pride transform into a planet the parcel of flesh and consciousness we are.

Man must vanquish himself, must do himself violence, in order to perform the slightest action untainted by evil.

To exist is equivalent to an act of faith, a protest against the truth, an interminable prayer....As soon as they consent to live, the unbeliever and the man of faith are fundamentally the same, since both have made the only decision that defines a *being*.

There is no means of proving it is preferable to be than not to be.

To want fame is to prefer dying scorned than forgotten.

Fear can supplant our real problems only to the extent—unwilling either to assimilate or to exhaust it—we perpetuate it within ourselves like a temptation and enthrone it at the very heart of our solitude.

What we want is not freedom but its appearances. It is for these simulacra that man has always striven. And since freedom, as has been said, is no more than a *sensation*, what difference is there between being free and believing ourselves free?

A civilization is destroyed only when its gods are destroyed.

To *exist* is a habit I do not despair of acquiring.

You are done for—a living dead man—not when you stop loving but stop hating. Hatred preserves: in it, in its chemistry, resides the "mystery" of life. Not for nothing is hatred still the best tonic ever discovered, for which any organism, however feeble, has a tolerance.

Afflicted with existence, each man endures like an animal the consequences which proceed from it. Thus, in a world where everything is detestable, hatred becomes huger than the world and, having transcended its object, cancels itself out.

Imaginary pains are by far the most real we suffer, since we feel a constant need for them and invent them because there is no way of doing without them.

One does not inhabit a country; one inhabits a language. That is our country, our fatherland—and no other.

Criticism is a misconception: we must read not to understand others but to understand ourselves.

No human being is more dangerous than those who have suffered for a belief: the great persecutors are recruited from the martyrs not quite beheaded. Far from diminishing the appetite for power, suffering exasperates it.

The fact that life has no meaning is a reason to live—moreover, the only one.

The mind is the result of the torments the flesh undergoes or inflicts upon itself.

Tyranny destroys or strengthens the individual; freedom enervates him, until he becomes no more than a puppet. Man has more chances of saving himself by hell than by paradise.

Sperm is a bandit in its pure state.

Reason is a whore, surviving by simulation, versatility, and shamelessness.

Negation is the mind's first freedom, yet a negative habit is fruitful only so long as we exert ourselves to overcome it, adapt it to our needs; once *acquired* it can imprison us.

When you have understood that nothing *is*, that things do not even deserve the status of appearances, you no longer need to be saved, you are saved, and miserable forever.

If we could see ourselves as others see us, we would vanish on the spot.

Much more than our other needs and endeavors, it is sexuality that puts us on an even footing with our kind: the more we practice it, the more we become like everyone else: it is in the performance of a reputedly bestial function that we prove our status as citizens: nothing is more *public* than the sexual act.

Speech and silence. We feel safer with a madman who talks than with one who cannot open his mouth.

We would not be interested in human beings if we did not have the hope of someday meeting someone worse off than ourselves.

Alone, even doing nothing, you do not waste your time. You do, almost always, in company. No encounter with yourself can be altogether sterile: Something necessarily emerges, even if only the hope of some day meeting yourself again.

It is not worth the bother of killing yourself, since you always kill yourself *too late.*

The obsession with suicide is characteristic of the man who can neither live nor die, and whose attention never swerves from this double impossibility.

Torment, for some men, is a need, an appetite, and an accomplishment.

We derive our vitality from our store of madness.

R. G. Collingwood

(1889-1943)
British philosopher

Parenthood is not an object of appetite or even desire. It is an object of will. There is no appetite for parenthood; there is only a purpose or intention of parenthood.

Like other revolutionaries I can thank God for the reactionaries. They clarify the issue.

What a man is ashamed of is always at bottom himself; and he is ashamed of himself at bottom always for being afraid.

Ralph J. Cudworth

(1617-88)
English theologian, philosopher

If intellection and knowledge were mere passion from without, or the bare reception of extraneous and adventitious forms, then no reason could be given at all why a mirror or looking-glass should not understand; whereas it cannot so much as sensibly perceive those images which it receives and reflects to us.

The true knowledge or science which exists nowhere but in the mind itself, has no other entity at all besides intelligibility; and therefore whatsoever is clearly intelligible, is absolutely true.

Knowledge is not a passion from without the mind, but an active exertion of the inward strength, vigour and power of the mind, displaying itself from within.

Sense is a line, the mind is a circle. Sense is like a line which is the flux of a point running out from itself, but intellect like a circle that keeps within itself.

Now, we deny not, but that politicians may sometimes abuse religion, and make it serve for the promoting of their own private interests and designs; which yet they could not do so well neither, were the thing itself a mere cheat and figment of their own, and had no reality at all in nature, nor anything solid at the bottom of it.

Truth is the most unbending and uncompliable, the most necessary, firm, immutable, and *adamantine* thing in the world.

John Dewey

(1859-1952)
American philosopher and educator

A democracy is more than a form of government; it is primarily a mode of associated living, of conjoint communicated experience.

The method of democracy is to bring conflicts out into the open where their special claims can be seen and appraised, where they can be discussed and judged.

Education is not preparation for life; education is life itself.

Rene Descartes

(1596-1650)
French philosopher and scientist

Cogito, ergo sum. (I think, therefore I am.)

Denis Diderot

(1713-84)
French philosopher

I have often seen an actor laugh off the stage, but I don't remember ever having seen one weep.

The following general definition of an animal: a system of different organic molecules that have combined with one another, under the impulsion of a sensation similar to an obtuse and muffled sense of touch given to them by the creator of matter as a whole, until each one of them has found the most suitable position for its shape and comfort.

To attempt the destruction of our passions is the height of folly. What a noble aim is that of the zealot who tortures himself like a madman in order to desire nothing, love nothing, feel nothing, and who, if he succeeded, would end up a complete monster!

It is said that desire is a product of the will, but the converse is in fact true: will is a product of desire.

The arbitrary rule of a just and enlightened prince is always bad. His virtues are the most dangerous and the surest form of seduction: they lull a people imperceptibly into the habit of loving, respecting, and serving his successor, whoever that successor may be, no matter how wicked or stupid.

Every man has his dignity. I'm willing to forget mine, but at my own discretion and not when someone else tells me to.

The possibility of divorce renders both marriage partners stricter in their observance of the duties they owe to each other. Divorces help to improve morals and to increase the population.

The most dangerous madmen are those created by religion, and...people whose aim is to disrupt society always know how to make good use of them on occasion.

No man has received from nature the right to give orders to others. Freedom is a gift from heaven, and every individual of the same species has the right to enjoy it as soon as he is in enjoyment of his reason.

The world is the house of the strong. I shall not know until the end what I have lost or won in this place, in this vast gambling den where I have spent more than sixty years, dicebox in hand, shaking the dice.

Genius is present in every age, but the men carrying it within them remain benumbed unless extraordinary events occur to heat up and melt the mass so that it flows forth.

Gaiety—a quality of ordinary men. Genius always presupposes some disorder in the machine.

It is not human nature we should accuse but the despicable conventions that pervert it.

The blood of Jesus Christ can cover a multitude of sins, it seems to me.

There are three principal means of acquiring knowledge available to us: observation of nature, reflection, and experimentation. Observation collects facts; reflection combines them; experimentation verifies the result of that combination. Our observation of nature must be diligent, our reflection profound, and our experiments exact. We rarely see these three means combined; and for this reason, creative geniuses are not common.

The general interest of the masses might take the place of the insight of genius if it were allowed freedom of action.

In any country where talent and virtue produce no advancement, money will be the national god. Its inhabitants will either have to possess money or make others believe that they do. Wealth will be the highest virtue, poverty the greatest vice. Those who have money will display it in every imaginable way. If their ostentation does not exceed their fortune, all will be well. But if their ostentation does exceed their fortune they will ruin themselves. In such a country, the greatest fortunes will vanish in the twinkling of an eye. Those who don't have money will ruin themselves with vain efforts to conceal their poverty. That is one kind of affluence: the outward sign of wealth for a small number, the mask of poverty for the majority, and a source of corruption for all.

Morals are in all countries the result of legislation and government; they are not African or Asian or European: they are good or bad.

Good music is very close to primitive language.

There is only one passion, the passion for happiness.

Patriotism is an ephemeral motive that scarcely ever outlasts the particular threat to society that aroused it.

The philosopher has never killed any priests, whereas the priest has killed a great many philosophers.

Poetry must have something in it that is barbaric, vast and wild.

When shall we see poets born? After a time of disasters and great misfortunes, when harrowed nations begin to breathe again. And then, shaken by the terror of such spectacles, imaginations will paint things entirely strange to those who have not witnessed them.

Disturbances in society are never more fearful than when those who are stirring up the trouble can use the pretext of religion to mask their true designs.

We are all instruments endowed with feeling and memory. Our senses are so many strings that are struck by surrounding objects and that also frequently strike themselves.

When superstition is allowed to perform the task of old age in dulling the human temperament, we can say goodbye to all excellence in poetry, in painting, and in music.

All abstract sciences are nothing but the study of relations between signs.

The pit of a theatre is the one place where the tears of virtuous and wicked men alike are mingled.

Only a very bad theologian would confuse the certainty that follows revelation with the truths that are revealed. They are entirely different things.

In order to shake a hypothesis, it is sometimes not necessary to do anything more than push it as far as it will go.

People praise virtue, but they hate it, they run away from it. It freezes you to death, and in this world you've got to keep your feet warm.

Diogenes of Sinope, The Cynic

(410 B.C.-320 B.C.)
Greek philosopher, moralist

Discourse on virtue and they pass by in droves, whistle and dance the shimmy, and you've got an audience.

Of what use is a philosopher who doesn't hurt anybody's feelings?

Why not whip the teacher when the pupil misbehaves?

In a rich man's house there is no place to spit but his face.

The art of being a slave is to rule one's master.

Ralph Waldo Emerson

(1803-82)
U.S. essayist, poet, philosopher

Men's actions are too strong for them. Show me a man who has acted, and who has not been the victim and slave of his action.

The German intellect wants the French sprightliness, the fine practical understanding of the English, and the American adventure; but it has a certain probity, which never rests in a superficial performance, but asks steadily, *To what end?* A German public asks for a controlling sincerity.

The thirst for adventure is the vent which Destiny offers; a war, a crusade, a gold mine, a new country, speak to the imagination and offer swing and play to the confined powers.

The moment we indulge our affections, the earth is metamorphosed; there is no winter and no night; all tragedies, all ennuis, vanish,—all duties even.

We do not quite forgive a giver. The hand that feeds us is in some danger of being bitten.

There is this to be said in favor of drinking, that it takes the drunkard first out of society, then out of the world.

Hitch your wagon to a star. Let us not fag in paltry works which serve our pot and bag alone.

The intellectual man requires a fine bait; the sots are easily amused. But everybody is drugged with his own frenzy, and the pageant marches at all hours, with music and banner and badge.

The angels are so enamoured of the language that is spoken in heaven, that they will not distort their lips with the hissing and unmusical dialects of men, but speak their own, whether there be any who understand it or not.

Who can…guess how much industry and providence and affection we have caught from the pantomime of brutes?

'Tis very certain that each man carries in his eye the exact indication of his rank in the immense scale of men, and we are always learning to read it. A complete man should need no auxiliaries to his personal presence.

I can reason down or deny everything, except this perpetual Belly: feed he must and will, and I cannot make him respectable.

The silence that accepts merit as the most natural thing in the world is the highest applause.

The aristocrat is the democrat ripe, and gone to seed.

If I cannot brag of knowing something, then I brag of not knowing it; at any rate, brag.

Each work of art excludes the world, concentrates attention on itself. For the time it is the only thing worth doing—to do just that; be it a sonnet, a statue, a landscape, an outline head of Caesar, or an oration. Presently we return to the sight of another that globes itself into a whole as did the first, for example, a beautiful garden; and nothing seems worth doing in life but laying out a garden.

Perpetual modernness is the measure of merit in every work of art.

Artists must be sacrificed to their art. Like bees, they must put their lives into the sting they give.

It is long ere we discover how rich we are. Our history, we are sure, is quite tame: we have nothing to write, nothing to infer. But our wiser years still run back to the despised recollections of childhood, and always we are fishing up some wonderful article out of that pond; until, by and by, we begin to suspect that the biography of the one foolish person we know is, in reality, nothing less than the miniature paraphrase of the hundred volumes of the Universal History.

Infancy conforms to nobody: all conform to it, so that one babe commonly makes four or five out of the adults who prattle and play to it.

We ascribe beauty to that which is simple; which has no superfluous parts; which exactly answers its end; which stands related to all things; which is the mean of many extremes.

Belief consists in accepting the affirmations of the soul; unbelief, in denying them.

We are born believing. A man bears beliefs as a tree bears apples.

The death of a dear friend, wife, brother, lover, which seemed nothing but privation, somewhat later assumes the aspect of a guide or genius; for it commonly operates revolutions in our way of life, terminates an epoch of infancy or of youth which was waiting to be closed, breaks up a wonted occupation, or a household, or style of living, and allows the formation of new ones more friendly to the growth of character.

There is properly no history, only biography.

Great geniuses have the shortest biographies.

We are too civil to books. For a few golden sentences we will turn over and actually read a volume of four or five hundred pages.

There are books…which take rank in your life with parents and lovers and passionate experiences, so medicinal, so stringent, so revolutionary, so authoritative.

I do not speak with any fondness but the language of coolest history, when I say that Boston commands attention as the town which was appointed in the destiny of nations to lead the civilization of North America.

There is also this benefit in brag, that the speaker is unconsciously expressing his own ideal. Humor him by all means, draw it all out, and hold him to it.

The right merchant is one who has the just average of faculties we call *common sense*; a man of a strong affinity for facts, who makes up his decision on what he has seen. He is thoroughly persuaded of the truths of arithmetic. There is always a reason, *in the man,* for his good or bad fortune…in making money. Men talk as if there were some magic about this….He knows that all goes on the old road, pound for pound, cent for cent—for every effect a perfect cause—and that good luck is another name for tenacity of purpose.

The attraction and superiority of California are in its days. It has better days & more of them, than any other country.

In skating over thin ice, our safety is in our speed.

Every burned book or house enlightens the world; every suppressed or expunged word reverberates through the earth from side to side.

A character is like an acrostic or Alexandrian stanza;—read it forward, backward, or across, it still spells the same thing.

Gross and obscure natures, however decorated, seem impure shambles; but character gives splendor to youth, and awe to wrinkled skin and gray hairs.

Do not tell me...of my obligation to put all poor men in good situations. Are they *my* poor? I tell thee, thou foolish philanthropist, that I grudge the dollar, the dime, the cent, I give to such men as do not belong to me and to whom I do not belong.

The child with his sweet pranks, the fool of his senses, commanded by every sight and sound, without any power to compare and rank his sensations, abandoned to a whistle or a painted chip, to a lead dragoon, or a gingerbread dog, individualizing everything, generalizing nothing, delighted with every new thing, lies down at night overpowered by the fatigue, which this day of continual pretty madness has incurred. But Nature has answered her purpose with the curly, dimpled lunatic. She has tasked every faculty, and has secured the symmetrical growth of the bodily frame, by all these attitudes and exertions—an end of the first importance, which could not be trusted to any care less perfect than her own.

Cities give us collision. 'Tis said, London and New York take the nonsense out of a man.

As long as our civilization is essentially one of property, of fences, of exclusiveness, it will be mocked by delusions. Our riches will leave us sick; there will be bitterness in our laughter; and our wine will burn our mouth. Only that good profits, which we can taste with all doors open, and which serves all men.

A right rule for a club would be, Admit no man whose presence excludes any one topic. It requires people who are not surprised and shocked, who do and let do, and let be, who sink trifles, and know solid values, and who take a great deal for granted.

The perception of the comic is a tie of sympathy with other men, a pledge of sanity, and a protection from those perverse tendencies and gloomy insanities in which fine intellects sometimes lose themselves. A rogue alive to the ludicrous is still convertible. If that sense is lost, his fellow-men can do little for him.

There is one topic peremptorily forbidden to all well-bred, to all rational mortals, namely, their distempers. If you have not slept, or if you have slept, or if you have headache, or sciatica, or leprosy, or thunder-stroke, I beseech you, by all angels, to hold your peace, and not pollute the morning.

Society everywhere is in conspiracy against the manhood of every one of its members....The virtue in most request is conformity. Self-reliance is its aversion. It loves not realities and creators, but names and customs.

One lesson we learn early, that in spite of seeming difference, men are all of one pattern. We readily assume this with our mates, and are disappointed and angry if we find that we are premature, and that their watches are slower than ours. In fact, the only sin which we never forgive in each other is difference of opinion.

All successful men have agreed in one thing,—they were *causationists*. They believed that things went not by luck, but by law; that there was not a weak or a cracked link in the chain that joins the first and last of things.

Men are conservatives when they are least vigorous, or when they are most luxurious. They are conservatives after dinner, or before taking their rest; when they are sick or aged. In the morning, or when their intellect or their conscience has been aroused, when they hear music, or when they read poetry, they are radicals.

All conservatives are such from personal defects. They have been effeminated by position or nature, born halt and blind, through luxury of their parents, and can only, like invalids, act on the defensive.

A foolish consistency is the hobgoblin of little minds, adored by little statesmen and philosophers and divines.

In every society some men are born to rule, and some to advise.

Every man is a consumer, and ought to be a producer....He is by constitution expensive, and needs to be rich.

Let me never fall into the vulgar mistake of dreaming that I am persecuted whenever I am contradicted.

In conversation the game is, to say something new with old words. And you shall observe a man of the people picking his way along, step by step, using every time an old boulder, yet never setting his foot on an old place.

Things said for conversation are chalk eggs. Don't *say* things. What you *are* stands over you the while, and thunders so that I cannot hear what you say to the contrary.

Courage charms us, because it indicates that a man loves an idea better than all things in the world, that he is thinking neither of his bed, nor his dinner, nor his money, but will venture all to put in act the invisible thought of his mind.

We must be as courteous to a man as we are to a picture, which we are willing to give the advantage of a good light.

It is the privilege of any human work which is well done to invest the doer with a certain haughtiness. He can well afford not to conciliate, whose faithful work will answer for him.

As men's prayers are a disease of the will, so are their creeds a disease of the intellect.

Commit a crime, and the earth is made of glass.

Men over forty are no judges of a book written in a new spirit.

It is said that the world is in a state of bankruptcy, that the world owes the world more than the world can pay.

Fate, then, is a name for facts not yet passed under the fire of thought; for causes which are unpenetrated.

The compensations of calamity are made apparent to the understanding also, after long intervals of time. A fever, a mutilation, a cruel disappointment, a loss of wealth, a loss of friends, seems at the moment unpaid loss, and unpayable. But the sure years reveal the deep remedial force that underlies all facts.

If a man knew anything, he would sit in a corner and be modest; but he is such an ignorant peacock, that he goes bustling up and down, and hits on extraordinary discoveries.

There are three wants which never can be satisfied: that of the rich, who wants something more; that of the sick, who wants something different; and that of the traveler, who says, "Anywhere but here."

I have heard with admiring submission the experience of the lady who declared that the sense of being perfectly well dressed gives a feeling of inward tranquillity which religion is powerless to bestow.

Tobacco and opium have broad backs, and will cheerfully carry the load of armies, if you choose to make them pay high for such joy as they give and such harm as they do.

Respect the child. Be not too much his parent. Trespass not on his solitude.

An empire is an immense egotism.

Coal is a portable climate. It carries the heat of the tropics to Labrador and the polar circle; and it is the means of transporting itself whither-soever it is wanted. Watt and Stephenson whispered in the ear of mankind their secret, that *a half-ounce of coal will draw two tons a mile*, and coal carries coal, by rail and by boat, to make Canada as warm as Calcutta, and with its comfort brings its industrial power.

I find the Englishman to be him of all men who stands firmest in his shoes. They have in themselves what they value in their horses, mettle and bottom.

Nothing great was ever achieved without enthusiasm.

Can we never extract the tapeworm of Europe from the brain of our countrymen?

We go to Europe to be Americanized.

'Tis a rule of manners to avoid exaggeration.

The world is upheld by the veracity of good men: they make the earth wholesome. They who lived with them found life glad and nutritious. Life is sweet and tolerable only in our belief in such society.

A man finds room in the few square inches of the face for the traits of all his ancestors; for the expression of all his history, and his wants.

Time dissipates to shining ether the solid angularity of facts.

Every fact is related on one side to sensation, and, on the other, to morals. The game of thought is, on the appearance of one of these two sides, to find the other; given the upper, to find the under side.

The first farmer was the first man, and all historic nobility rests on possession and use of land.

A man's personal defects will commonly have with the rest of the world precisely that importance which they have to himself. If he makes light of them, so will other men.

He has not learned the lesson of life who does not every day surmount a fear.

We estimate the wisdom of nations by seeing what they did with their surplus capital.

Flowers...are a proud assertion that a ray of beauty outvalues all the utilities of the world.

Let the stoics say what they please, we do not eat for the good of living, but because the meat is savory and the appetite is keen.

A friend may well be reckoned the masterpiece of Nature.

I do then with my friends as I do with my books. I would have them where I can find them, but I seldom use them.

It is always so pleasant to be generous, though very vexatious to pay debts.

In every work of genius we recognize our own rejected thoughts: they come back to us with a certain alienated majesty.

The hearing ear is always found close to the speaking tongue; and no genius can long or often utter anything which is not invited and gladly entertained by men around him.

Repose and cheerfulness are the badge of the gentleman,—repose in energy.

The only gift is a portion of thyself.

The dice of God are always loaded.

'Tis the old secret of the gods that they come in low disguises.

Let us treat the men and women well: treat them as if they were real: perhaps they are.

To be great is to be misunderstood.

The search after the great men is the dream of youth, and the most serious occupation of manhood.

Bad times have a scientific value. These are occasions a good learner would not miss.

Heroism feels and never reasons, and therefore is always right.

Every hero becomes a bore at last.

Nothing astonishes men so much as common sense and plain dealing.

The louder he talked of his honor, the faster we counted our spoons.

What is the imagination? Only an arm or weapon of the interior energy; only the precursor of the reason.

Higher than the question of our duration is the question of our deserving. Immortality will come to such as are fit for it, and he would be a great soul in future must be a great soul now.

Wise men are not wise at all hours, and will speak five times from their taste or their humor, to once from their reason.

Who shall set a limit to the influence of a human being?

Of course, money will do after its kind, and will steadily work to unspiritualize and unchurch the people to whom it was bequeathed.

The torpid artist seeks inspiration at any cost, by virtue or by vice, by friend or by fiend, by prayer or by wine.

An institution is the lengthened shadow of one man.

One definition of man is "an intelligence served by organs."

Nations have lost their old omnipotence; the patriotic tie does not hold. Nations are getting obsolete, we go and live where we will.

Man is a shrewd inventor, and is ever taking the hint of a new machine from his own structure, adapting some secret of his own anatomy in iron, wood, and leather, to some required function in the work of the world.

The world is full of judgment-days, and into every assembly that a man enters, in every action he attempts, he is gauged and stamped.

If a man owns land, the land owns him.

Language is the archives of history.

I like to be beholden to the great metropolitan English speech, the sea which receives tributaries from every region under heaven.

The wise know that foolish legislation is a rope of sand, which perishes in the twisting.

The good lawyer is not the man who has an eye to every side and angle of contingency, and qualifies all his qualifications, but who throws himself on your part so heartily, that he can get you out of a scrape.

The studious class are their own victims: they are thin and pale, their feet are cold, their heads are hot, the night is without sleep, the day a fear of interruption—pallor, squalor, hunger, and egotism.

Meek young men grow up in libraries, believing it their duty to accept the views which Cicero, which Locke, which Bacon, have given, forgetful that Cicero, Locke, and Bacon were only young men in libraries, when they wrote these books. Hence, instead of Man Thinking, we have the book-worm.

A man's library is a sort of harem.

Be a little careful about your library. Do you foresee what you will do with it? Very little to be sure. But the real question is, What it will do with you? You will come here and get books that will open your eyes, and your ears, and your curiosity, and turn you inside out or outside in.

Every violation of truth is not only a sort of suicide in the liar, but is a stab at the health of human society.

There is then creative reading as well as creative writing. When the mind is braced by labor and invention, the page of whatever book we read becomes luminous with manifold allusion. Every sentence is doubly significant, and the sense of our author is as broad as the world.

The best bribe which London offers to-day to the imagination, is, that, in such a vast variety of people and conditions, one can believe there is room for persons of romantic character to exist, and that the poet, the mystic, and the hero may hope to confront their counterparts.

He who is in love is wise and is becoming wiser, sees newly every time he looks at the object beloved, drawing from it with his eyes and his mind those virtues which it possesses.

All mankind love a lover.

There is no chance, and no anarchy, in the universe. All is system and gradation. Every god is there sitting in his sphere.

By his machines man can dive and remain under water like a shark; can fly like a hawk in the air; can see atoms like a gnat; can see the system of the universe of Uriel, the angel of the sun; can carry whatever loads a ton of coal can lift; can knock down cities with his fist of gunpowder; can recover the history of his race by the medals which the deluge, and every creature, civil or savage or brute, has involuntarily dropped of its existence; and divine the future possibility of the planet and its inhabitants by his perception of laws of nature.

Shall we then judge a country by the majority, or by the minority?

There are men whose manners have the same essential splendor as the simple and awful sculpture on the friezes of the Parthenon, and the remains of the earliest Greek art.

Manners are the happy way of doing things; each once a stroke of genius or of love—now repeated and hardened into usage. They form at last a rich varnish, with which the routine of life is washed, and its details adorned. If they are superficial, so are the dewdrops which give such depth to the morning meadows.

Good manners are made up of petty sacrifices.

The martyr cannot be dishonored. Every lash inflicted is a tongue of fame; every prison a more illustrious abode.

The torments of martyrdom are probably most keenly felt by the bystanders.

Leave this hypocritical prating about the masses. Masses are rude, lame, unmade, pernicious in their demands and influence, and need not to be flattered, but to be schooled. I wish not to concede anything to them, but to tame, drill, divide, and break them up, and draw individuals out of them.

All history is a record of the power of minorities, and of minorities of one.

The mob is man voluntarily descending to the nature of the beast. Its fit hour of activity is night. Its actions are insane like its whole constitution. It persecutes a principle; it would whip a right; it would tar and feather justice, by inflicting fire and outrage upon the houses and persons of those who have these. It resembles the prank of boys, who run with fire-engines to put out the ruddy aurora streaming to the stars.

Money, which represents the prose of life, and which is hardly spoken of in parlors without an apology, is, in its effects and laws, as beautiful as roses.

Murder in the murderer is no such ruinous thought as poets and romancers will have it; it does not unsettle him, or fright him from his ordinary notice of trifles; it is an act quite easy to be contemplated.

Solvency is maintained by means of a national debt, on the principle, "If you will not lend me the money, how can I pay you?"

Shall we then judge a country by the majority, or by the minority? By the minority, surely. 'Tis pedantry to estimate nations by the census, or by square miles of land, or other than by their importance to the mind of the time.

Nature is an endless combination and repetition of a very few laws. She hums the old well-known air through innumerable variations.

The wonder is always new that any sane man can be a sailor.

The most advanced nations are always those who navigate the most.

We do what we must, and call it by the best names.

As every man is hunted by his own demon, vexed by his own disease, this checks all his activity.

What forests of laurel we bring, and the tears of mankind, to those who stood firm against the opinion of their contemporaries!

The reason why men do not obey us is because they see the mud at the bottom of our eye.

We do not count a man's years until he has nothing else to count.

Twenty thousand thieves landed at Hastings. These founders of the House of Lords were greedy and ferocious dragoons, sons of greedy and ferocious pirates. They were all alike, they took everything they could carry, they burned, harried, violated, tortured, and killed until everything English was brought to the verge of ruin. Such, however, is the illusion of antiquity and wealth, that decent and dignified men now existing boast their descent from these filthy thieves, who showed a far juster conviction of their own merits, by assuming for their types the swine, goat, jackal, leopard, wolf, and snake, which they severally resembled.

Passion, though a bad regulator, is a powerful spring.

The history of persecution is a history of endeavors to cheat nature, to make water run up hill, to twist a rope of sand.

The measure of a master is his success in bringing all men round to his opinion twenty years later.

Only poetry inspires poetry.

A sect or a party is an elegant incognito, devised to save a man from the vexation of thinking.

If government knew how, I should like to see it check, not multiply, the population. When it reaches its true law of action, every man that is born will be hailed as essential.

Some men are born to own, and can animate all their possessions. Others cannot: their owning is not graceful; seems to be a compromise of their character: they seem to steal their own dividends.

Give me insight into today and you may have the antique and future worlds.

The President has paid dear for his White House. It has commonly cost him all his peace, and the best of his manly attributes. To preserve for a short time so conspicuous an appearance before the world, he is content to eat dust before the real masters who stand erect behind the throne.

No man acquires property without acquiring with it a little arithmetic also.

Property is an intellectual production. The game requires coolness, right reasoning, promptness, and patience in the players.

The Anglican Church is marked by the grace and good sense of its forms, by the manly grace of its clergy. The gospel it preaches is, "By taste are ye saved."…It is not in ordinary a persecuting church; it is not inquisitorial, not even inquisitive, is perfectly well bred and can shut its eyes on all proper occasions. If you let it alone, it will let you alone. But its instinct is hostile to all change in politics, literature, or social arts.

I hate quotations. Tell me what you know.

Next to the originator of a good sentence is the first quoter of it. Many will read the book before one thinks of quoting a passage. As soon as he has done this, that line will be quoted east and west.

By necessity, by proclivity, and by delight, we all quote.

The spirit of our American radicalism is destructive and aimless; it is not loving; it has no ulterior and divine ends; but is destructive only out of hatred and selfishness.

Never read any book that is not a year old.

'Tis the good reader that makes the good book; a good head cannot read amiss: in every book he finds passages which seem confidences or asides hidden from all else and unmistakably meant for his ear.

Can anybody remember when the times were not hard, and money not scarce?

Every reform was once a private opinion, and when it shall be a private opinion again, it will solve the problem of the age.

Dear to us are those who love us…but dearer are those who reject us as unworthy, for they add another life; they build a heaven before us whereof we had not dreamed, and thereby supply to us new powers out of the recesses of the spirit, and urge us to new and unattempted performances.

Trust men, and they will be true to you; treat them greatly, and they will show themselves great.

If there is any period one would desire to be born in, is it not the age of Revolution; when the old and the new stand side by side, and admit of being compared; when the energies of all men are searched by fear and by hope; when the historic glories of the old can be compensated by the rich possibilities of the new era?

The office of the scholar is to cheer, to raise, and to guide men by showing them facts amidst appearances. He plies the slow, unhonored, and unpaid task of observation....He is the world's eye.

What terrible questions we are learning to ask! The former men believed in magic, by which temples, cities, and men were swallowed up, and all trace of them gone. We are coming on the secret of a magic which sweeps out of men's minds all vestige of theism and beliefs which they and their fathers held and were framed upon.

The sea, washing the equator and the poles, offers its perilous aid, and the power and empire that follow it...."Beware of me," it says, "but if you can hold me, I am the key to all the lands."

Trust thyself: every heart vibrates to that iron string.

Welcome evermore to gods and men is the self-helping man. For him all doors are flung wide: him all tongues greet, all honors crown, all eyes follow with desire. Our love goes out to him and embraces him, because he did not need it. We solicitously and apologetically caress and celebrate him, because he held on his way and scorned our disapprobation. The gods loved him because men hated him.

Let a man then know his worth, and keep things under his feet. Let him not peep or steal, or skulk up and down with the air of a charity-boy, a bastard, or an interloper.

Sentimentalists…adopt whatever merit is in good repute, and almost make it hateful with their praise. The warmer their expressions, the colder we feel.…Cure the drunkard, heal the insane, mollify the homicide, civilize the Pawnee, but what lessons can be devised for the debauchee of sentiment?

That which we call sin in others, is experiment for us.

Sincerity is the luxury allowed, like diadems and authority, only to the highest rank.…Every man alone is sincere. At the entrance of a second person, hypocrisy begins.

The sky is the daily bread of the eyes.

Society never advances. It recedes as fast on one side as it gains on the other.…Society acquires new arts, and loses old instincts.

Society is a masked ball, where every one hides his real character, and reveals it by hiding.

Sorrow makes us all children again, destroys all differences of intellect. The wisest knows nothing.

Speech is power: speech is to persuade, to convert, to compel. It is to bring another out of his bad sense into your good sense.

Our spontaneous action is always the best. You cannot, with your best deliberation and heed, come so close to any question as your spontaneous glance shall bring you.

The State must follow, and not lead, the character and progress of the citizen.

If a man can write a better book, preach a better sermon, or make a better mouse-trap, than his neighbor, though he build his house in the woods, the world will make a beaten path to his door.

A man is known by the books he reads, by the company he keeps, by the praise he gives, by his dress, by his tastes, by his distastes, by the stories he tells, by his gait, by the notion of his eye, by the look of his house, of his chamber; for nothing on earth is solitary but every thing hath affinities infinite.

We boil at different degrees.

To think is to act.

The revelation of Thought takes men out of servitude into freedom.

The surest poison is time.

We rail at trade, but the historian of the world will see that it was the principle of liberty; that it settled America, and destroyed feudalism, and made peace and keeps peace; that it will abolish slavery.

The greatest meliorator of the world is selfish, huckstering Trade.

I do not hesitate to read...all good books in translations. What is really best in any book is translatable—any real insight or broad human sentiment.

Traveling is a fool's paradise. Our first journeys discover to us the indifference of places.

I am not much an advocate for traveling, and I observe that men run away to other countries because they are not good in their own, and run back to their own because they pass for nothing in the new places. For the most part, only the light characters travel. Who are you that have no task to keep you at home?

The highest compact we can make with our fellow is—"Let there be truth between us two forevermore."

God offers to every mind its choice between truth and repose. Take which you please; you can never have both.

The secret of ugliness consists not in irregularity, but in being uninteresting.

I hate this shallow Americanism which hopes to get rich by credit, to get knowledge by raps on midnight tables, to learn the economy of the mind by phrenology, or skill without study, or mastery without apprenticeship.

We are a puny and fickle folk. Avarice, hesitation, and following are our diseases.

In America the geography is sublime, but the men are not; the inventions are excellent, but the inventors one is sometimes ashamed of.

I suffer whenever I see that common sight of a parent or senior imposing his opinion and way of thinking and being on a young soul to which they are totally unfit. Cannot we let people be themselves, and enjoy life in their own way? You are trying to make that man another you. One's enough.

As there is a use in medicine for poisons, so the world cannot move without rogues.

The less a man thinks or knows about his virtues, the better we like him.

The virtues of society are vices of the saint. The terror of reform is the discovery that we must cast away our virtues, or what we have always esteemed such, into the same pit that has consumed our grosser vices.

The triumphs of peace have been in some proximity to war. Whilst the hand was still familiar with the sword-hilt, whilst the habits of the camp were still visible in the port and complexion of the gentleman, his intellectual power culminated; the compression and tension of these stern conditions is a training for the finest and softest arts, and can rarely be compensated in tranquil times, except by some analogous vigor drawn from occupations as hardy as war.

Wealth is in applications of mind to nature; and the art of getting rich consists not in industry, much less in saving, but in a better order, in timeliness, in being at the right spot.

I greet you at the beginning of a great career, which must yet have had a long foreground somewhere, for such a start. I rubbed my eyes a little to see if this sunbeam were no illusion; but the solid sense of the book is a sober certainty. It has the best merits, namely, of fortifying and encouraging.

Raphael paints wisdom; Handel sings it, Phidias carves it, Shakespeare writes it, Wren builds it, Columbus sails it, Luther preaches it, Washington arms it, Watt mechanizes it.

Wisdom is like electricity. There is no permanently wise man, but men capable of wisdom, who, being put into certain company, or other favorable conditions, become wise for a short time, as glasses rubbed acquire electric power for a while.

Slavery it is that makes slavery; freedom, freedom. The slavery of women happened when the men were slaves of kings.

To judge from a single conversation, he made the impression of a narrow and very English mind; of one who paid for his rare elevation by general tameness and conformity. Off his own beat, his opinions were of no value.

We must hold a man amenable to reason for the choice of his daily craft or profession. It is not an excuse any longer for his deeds that they are the custom of his trade. What business has he with an evil trade?

The mark of the man of the world is absence of pretension. He does not make a speech; he takes a low business-tone, avoids all brag, is nobody, dresses plainly, promises not at all, performs much, speaks in monosyllables, hugs his fact. He calls his employment by its lowest name, and so takes from evil tongues their sharpest weapon. His conversation clings to the weather and the news, yet he allows himself to be surprised into thought, and the unlocking of his learning and philosophy.

There is no luck in literary reputation. They who make up the final verdict upon every book are not the partial and noisy readers of the hour when it appears; but a court as of angels, a public not to be bribed, not to be entreated, and not to be overawed, decides upon every man's title to fame.

Talent alone cannot make a writer. There must be a man behind the book; a personality which, by birth and quality, is pledged to the doctrines there set forth, and which exists to see and state things so, and not otherwise.

Friedrich Engels

(1820-95)
German social philosopher

An ounce of action is worth a ton of theory.

By bourgeoisie is meant the class of modern capitalists, owners of the means of social production and employers of wage labor. By proletariat, the class of modern wage laborers who, having no means of production of their own, are reduced to selling their labor power in order to live.

People think they have taken quite an extraordinarily bold step forward when they have rid themselves of belief in hereditary monarchy and swear by the democratic republic. In reality, however, the state is nothing but a machine for the oppression of one class by another, and indeed in the democratic republic no less than in the monarchy.

Epictetus

(60 B.C-?)
Greek Stoic philosopher

It is a sign of a dull nature to occupy oneself deeply in matters that concern the body; for instance, to be over much occupied about exercise, about eating and drinking, about easing oneself, about sexual intercourse.

Remember that you are an actor in a drama, of such a part as it may please the master to assign you, for a long time or for a little as he may choose. And if he will you to take the part of a poor man, or a cripple, or a ruler, or a private citizen, then may you act that part with grace! For to act well the part that is allotted to us, that indeed is ours to do, but to choose it is another's.

You are a little soul carrying around a corpse.

Never in any case say *I have lost* such a thing, but *I have returned it.* Is your child dead? It is a return. Is your wife dead? It is a return. Are you deprived of your estate? Is not this also a return?

If you set your heart upon philosophy, you must straightway prepare yourself to be laughed at and mocked by many who will say *Behold a philosopher arisen among us!* or *How came you by that brow of scorn?* But do you cherish no scorn, but hold to those things which seem to you the best, as one set by God in that place. Remember too, that if you abide in those ways, those who first mocked you, the same shall afterwards reverence you; but if you yield to them, you will be laughed at twice as much as before.

You may be always victorious if you will never enter into any contest where the issue does not wholly depend upon yourself.

I have never wished to cater to the crowd; for what I know they do not approve, and what they approve I do not know.

Frantz Fanon

(1925-61)
Martiniquan psychiatrist, philosopher, political activist

However painful it may be for me to accept this conclusion, I am obliged to state it: for the black man there is only one destiny. And it is white.

I am black: I am the incarnation of a complete fusion with the world, an intuitive understanding of the earth, an abandonment of my ego in the heart of the cosmos, and no white man, no matter how intelligent he may be, can ever understand Louis Armstrong and the music of the Congo.

Fervor is the weapon of choice of the impotent.

I ascribe a basic importance to the phenomenon of language....To speak means to be in a position to use a certain syntax, to grasp the morphology of this or that language, but it means above all to assume a culture, to support the weight of a civilization.

There is a point at which methods devour themselves.

What I call middle-class society is any society that becomes rigidified in predetermined forms, forbidding all evolution, all gains, all progress, all discovery. I call middle-class a closed society in which life has no taste, in which the air is tainted, in which ideas and men are corrupt. And I think that a man who takes a stand against this death is in a sense a revolutionary.

Collective guilt is borne by what is conventionally called the scapegoat. Now the scapegoat for white society—which is based on myths of progress, civilization, liberalism, education, enlightenment, refinement—will be precisely the force that opposes the expansion and the triumph of these myths. This brutal opposing force is supplied by the Negro.

Ludwig Feuerbach

(1804-72)
German philosopher

If therefore my work is negative, irreligious, atheistic, let it be remembered that atheism—at least in the sense of this work—is the secret of religion itself; that religion itself, not indeed on the surface, but fundamentally, not in intention or according to its own supposition, but in its heart, in its essence, believes in nothing else than the truth and divinity of human nature.

I have always taken as the standard of the mode of teaching and writing, not the abstract, particular, professional philosopher, but universal man, that I have regarded *man* as the criterion of truth, and not this or that founder of a system, and have from the first placed the highest excellence of the philosopher in this, that he abstains, both as a man and as an author, from the ostentation of philosophy, *i.e.*, that he is a philosopher only in reality, not formally, that he is a quiet philosopher, not a loud and still less a brawling one.

Religion is the dream of the human mind. But even in dreams we do not find ourselves in emptiness or in heaven, but on earth, in the realm of reality; we only see real things in the entrancing splendor of imagination and caprice, instead of in the simple daylight of reality and necessity.

The present age...prefers the sign to the thing signified, the copy to the original, fancy to reality, the appearance to the essence...for in these days *illusion* only is *sacred, truth profane.*

Michel Foucault

(1926-1984)
French philosopher

Freedom of conscience entails more dangers than authority and despotism.

The lyricism of marginality may find inspiration in the image of the "outlaw," the great social nomad, who prowls on the confines of a docile, frightened order.

The strategic adversary is fascism...the fascism in us all, in our heads and in our everyday behavior, the fascism that causes us to love power, to desire the very thing that dominates and exploits us.

There is a sort of myth of History that philosophers have....History for philosophers is some sort of great, vast continuity in which the freedom of individuals and economic or social determinations come and get entangled. When someone lays a finger on one of those great themes— continuity, the effective exercise of human liberty, how individual liberty is articulated with social determinations—when someone touches one of these three myths, these good people start crying out that History is being raped or murdered.

Sexuality is a part of our behavior. It's part of our world freedom. Sexuality is something that we ourselves create. It is our own creation, and much more than the discovery of a secret side of our desire. We have to understand that with our desires go new forms of relationships, new forms of love, new forms of creation. Sex is not a fatality; it's a possibility for creative life. It's not enough to affirm that we are gay but we must also create a gay life.

As the archeology of our thought easily shows, man is an invention of recent date. And one perhaps nearing its end.

There are more ideas on earth than intellectuals imagine. And these ideas are more active, stronger, more resistant, more passionate than "politicians" think. We have to be there at the birth of ideas, the bursting outward of their force: not in books expressing them, but in events manifesting this force, in struggles carried on around ideas, for or against them. Ideas do not rule the world. But it is because the world has ideas…that it is not passively ruled by those who are its leaders or those who would like to teach it, once and for all, what it must think.

The work of an intellectual is not to mould the political will of others; it is, through the analyses that he does in his own field, to re-examine evidence and assumptions, to shake up habitual ways of working and thinking, to dissipate conventional familiarities, to re-evaluate rules and institutions and...to participate in the formation of a political will (where he has his role as citizen to play).

The judges of normality are present everywhere. We are in the society of the teacher-judge, the doctor-judge, the educator-judge, the "social worker"-judge.

Justice must always question itself, just as society can exist only by means of the work it does on itself and on its institutions.

Chance does not speak essentially through words nor can it be seen in their convolution. It is the eruption of language, its sudden appearance....It's not a night atwinkle with stars, an illuminated sleep, nor a drowsy vigil. It is the very edge of consciousness.

Madness is the absolute break with the work of art; it forms the constitutive moment of abolition, which dissolves in time the truth of the work of art.

During the years 1945-1965 (I am referring to Europe), there was a certain way of thinking correctly, a certain style of political discourse, a certain ethics of the intellectual. One had to be on familiar terms with Marx, not let one's dreams stray too far from Freud....These were the...requirements that made the strange occupation of writing and speaking a measure of truth about oneself and one's time acceptable.

Power is not an institution, and not a structure; neither is it a certain strength we are endowed with; it is the name that one attributes to a complex strategical situation in a particular society.

Prison continues, on those who are entrusted to it, a work begun elsewhere, which the whole of society pursues on each individual through innumerable mechanisms of discipline.

Psychoanalysis can unravel some of the forms of madness; it remains a stranger to the sovereign enterprise of unreason. It can neither limit nor transcribe, nor most certainly explain, what is essential in this enterprise.

In its function, the power to punish is not essentially different from that of curing or educating.

If repression has indeed been the fundamental link between power, knowledge, and sexuality since the classical age, it stands to reason that we will not be able to free ourselves from it except at a considerable cost.

Georg Hegel

(1770-1831)
German philosopher

The first glance at History convinces us that the actions of men proceed from their needs, their passions, their characters and talents; and impresses us with the belief that such needs, passions and interests are the sole spring of actions.

Animals are in possession of themselves; their soul is in possession of their body. But they have no right to their life, because they do not will it.

The true courage of civilized nations is readiness for sacrifice in the service of the state, so that the individual counts as only one amongst many. The important thing here is not personal mettle but aligning oneself with the universal.

The true theater of history is therefore the temperate zone.

In history an additional result is commonly produced by human actions beyond that which they aim at and obtain—that which they immediately recognize and desire. They gratify their own interest; but something further is thereby accomplished, latent in the actions in question, though not present to their consciousness, and not included in their design.

Education is the art of making man ethical.

We do not need to be shoemakers to know if our shoes fit, and just as little have we any need to be professionals to acquire knowledge of matters of universal interest.

It is easier to discover a deficiency in individuals, in states, and in Providence, than to see their real import and value.

The history of the world is none other than the progress of the consciousness of freedom.

An idea is always a generalization, and generalization is a property of thinking. To generalize means to think.

Mere goodness can achieve little against the power of nature.

The History of the world is not the theatre of happiness. Periods of happiness are blank pages in it, for they are periods of harmony—periods when the antithesis is in abeyance.

Once the state has been founded, there can no longer be any heroes. They come on the scene only in uncivilized conditions.

But what experience and history teach is this—that peoples and governments have never learned anything from history, or acted on principles deduced from it.

World history is a court of judgment.

Regarding History as the slaughter-bench at which the happiness of peoples, the wisdom of States, and the virtue of individuals have been victimized—the question involuntarily arises—to what principle, to what final aim these enormous sacrifices have been offered.

When liberty is mentioned, we must always be careful to observe whether it is not really the assertion of private interests which is thereby designated.

When we walk the streets at night in safety, it does not strike us that this might be otherwise. This habit of feeling safe has become second nature, and we do not reflect on just how this is due solely to the working of special institutions. Commonplace thinking often has the impression that force holds the state together, but in fact its only bond is the fundamental sense of order which everybody possesses.

It is a matter of perfect indifference where a thing originated; the only question is: "Is it true in and for itself?"

Truth in philosophy means that concept and external reality correspond.

Poverty in itself does not make men into a rabble; a rabble is created only when there is joined to poverty a disposition of mind, an inner indignation against the rich, against society, against the government.

Amid the pressure.of great events, a general principle gives no help.

Public opinion contains all kinds of falsity and truth, but it takes a great man to find the truth in it. The great man of the age is the one who can put into words the will of his age, tell his age what its will is, and accomplish it. What he does is the heart and the essence of his age, he actualizes his age. The man who lacks sense enough to despise public opinion expressed in gossip will never do anything great.

To him who looks upon the world rationally, the world in its turn presents a rational aspect. The relation is mutual.

The learner always begins by finding fault, but the scholar sees the positive merit in everything.

As high as mind stands above nature, so high does the state stand above physical life. Man must therefore venerate the state as a secular deity....The march of God in the world, that is what the State is.

Martin Heidegger

(1889-1976)
German philosopher

The German language "speaks Being," while all the others merely "speak of Being."

Man acts as though he were the shaper and master of language, while in fact language remains the master of man.

Heraclitus

(c.535-c.475 B.C.)
Greek philosopher

Corpses are more fit to be thrown out than is dung.

It is hard to contend against one's heart's desire; for whatever it wishes to have it buys at the cost of soul.

To do the same thing over and over again is not only boredom: it is to be controlled by rather than to control what you do.

You could not step twice into the same rivers; for other waters are ever flowing on to you.

Change alone is unchanging.

Men who wish to know about the world must learn about it in its particular details.

The world, an entity out of everything, was created by neither gods nor men, but was, is and will be eternally living fire, regularly becoming ignited and regularly becoming extinguished.

God is day and night, winter and summer, war and peace, surfeit and hunger.

History is a child building a sand-castle by the sea, and that child is the whole majesty of man's power in the world.

Hide our ignorance as we will, an evening of wine soon reveals it.

Immortal mortals, mortal immortals, one living the others' death and dying the others' life.

Even sleepers are workers and collaborators on what goes on in the universe.

Thomas Hobbes

(1588-1679)
English philosopher

The privilege of absurdity; to which no living creature is subject, but man only.

War consisteth not in battle only, or the act of fighting; but in a tract of time, wherein the will to contend by battle is sufficiently known.

The praise of ancient authors proceeds not from the reverence of the dead, but from the competition and mutual envy of the living.

A man's conscience and his judgment is the same thing; and as the judgment, so also the conscience, may be erroneous.

Man is distinguished, not only by his reason; but also by this singular passion from other animals...which is a lust of the mind, that by a perseverance of delight in the continual and indefatigable generation of knowledge, exceeds the short vehemence of any carnal pleasure.

Leisure is the mother of Philosophy.

The Papacy is no other than the ghost of the deceased Roman empire, sitting crowned upon the grave thereof.

No arts; no letters; no society; and which is worst of all, continual fear, and danger of violent death; and the life of man, solitary, poor, nasty, brutish, and short.

Wherein men live without other security, than what their own strength and their own invention shall furnish them.

He that is taken and put into prison or chains is not conquered, though overcome; for he is still an enemy.

For it is with the mysteries of our religion, as with wholesome pills for the sick, which swallowed whole, have the virtue to cure; but chewed, are for the most part cast up again without effect.

Science is the knowledge of consequences, and dependence of one fact upon another.

The secret thoughts of a man run over all things, holy, profane, clean, obscene, grave, and light, without shame or blame.

The obligation of subjects to the sovereign is understood to last as long, and no longer, than the power lasteth by which he is able to protect them.

There is no such thing as perpetual tranquillity of mind while we live here; because life itself is but motion, and can never be without desire, nor without fear, no more than without sense.

Such truth, as opposeth no man's profit, nor pleasure, is to all men welcome.

Such is the nature of men, that howsoever they may acknowledge many others to be more witty, or more eloquent, or more learned; yet they will hardly believe there be many so wise as themselves.

Eric Hoffer

(1902-83)
U.S. philosopher

Action is at bottom a swinging and flailing of the arms to regain one's balance and keep afloat.

One of the marks of a truly vigorous society is the ability to dispense with passion as a midwife of action—the ability to pass directly from thought to action.

Perhaps a modern society can remain stable only by eliminating adolescence, by giving its young, from the age of ten, the skills, responsibilities, and rewards of grownups, and opportunities for action in all spheres of life. Adolescence should be a time of useful action, while book learning and scholarship should be a preoccupation of adults.

Our greatest pretenses are built up not to hide the evil and the ugly in us, but our emptiness. The hardest thing to hide is something that is not there.

Man is the only creature that strives to surpass himself, and yearns for the impossible.

Animals often strike us as passionate machines.

The remarkable thing is that it is the crowded life that is most easily remembered. A life full of turns, achievements, disappointments, surprises, and crises is a life full of landmarks. The empty life has even its few details blurred, and cannot be remembered with certainty.

The difficult and risky task of meeting and mastering the new—whether it be the settlement of new lands or the initiation of new ways of life—is not undertaken by the vanguard of society but by its rear. It is the misfits, failures, fugitives, outcasts and their like who are among the first to grapple with the new.

When we believe ourselves in possession of the only truth, we are likely to be indifferent to common everyday truths.

Capitalism is at its liberating best in a noncapitalist environment. The crypto-businessman is the true revolutionary in a Communist country.

Unpredictability, too, can become monotonous.

There are no chaste minds. Minds copulate wherever they meet.

Compassion is the antitoxin of the soul: where there is compassion even the most poisonous impulses remain relatively harmless.

"More!" is as effective a revolutionary slogan as was ever invented by doctrinaires of discontent. The American, who cannot learn to want what he has, is a permanent revolutionary.

What greater reassurance can the weak have than that they are like anyone else?

When cowardice is made respectable, its followers are without number both from among the weak and the strong; it easily becomes a fashion.

Our credulity is greatest concerning the things we know least about. And since we know least about ourselves, we are ready to believe all that is said about us. Hence the mysterious power of both flattery and calumny.

It is thus with most of us: we are what other people say we are. We know ourselves chiefly by hearsay.

The world leans on us. When we sag, the whole world seems to droop.

Disappointment is a sort of bankruptcy—the bankruptcy of a soul that expends too much in hope and expectation.

The chemistry of dissatisfaction is as the chemistry of some marvelously potent tar. In it are the building stones of explosives, stimulants, poisons, opiates, perfumes and stenches.

The beginning of thought is in disagreement—not only with others but also with ourselves.

Though dissenters seem to question everything in sight, they are actually bundles of dusty answers and never conceived a new question.

What offends us most in the literature of dissent is the lack of hesitation and wonder.

Dissipation is a form of self-sacrifice.

We do not really feel grateful toward those who make our dreams come true; they ruin our dreams.

We have perhaps a natural fear of ends. We would rather be always on the way than arrive. Given the means, we hang on to them and often forget the ends.

You can discover what your enemy fears most by observing the means he uses to frighten you.

Man staggers through life yapped at by his reason, pulled and shoved by his appetites, whispered to by fears, beckoned by hopes. Small wonder that what he craves most is self-forgetting.

The prehuman creature from which man evolved was unlike any other living thing in its malicious viciousness toward its own kind.... Humanization was not a leap forward but a groping toward survival.

Thought is a process of exaggeration. The refusal to exaggerate is not infrequently an alibi for the disinclination to think or praise.

It is not so much the example of others we imitate as the reflection of ourselves in their eyes and the echo of ourselves in their words.

The individual who has to justify his existence by his own efforts is in eternal bondage to himself.

Facts are counterrevolutionary.

There is no loneliness greater than the loneliness of a failure. The failure is a stranger in his own house.

Our achievements speak for themselves. What we have to keep track of are our failures, discouragements, and doubts. We tend to forget the past difficulties, the many false starts, and the painful groping. We see our past achievements as the end result of a clean forward thrust, and our present difficulties as signs of decline and decay.

Absolute faith corrupts as absolutely as absolute power.

Men weary as much of not doing the things they want to do as of doing the things they do not want to do.

The pleasure we derive from doing favors is partly in the feeling it gives us that we are not altogether worthless. It is a pleasant surprise to ourselves.

There is always a chance that he who sets himself up as his brother's keeper will end up by being his jailkeeper.

The basic test of freedom is perhaps less in what we are free to do than in what we are free not to do. It is the freedom to refrain, withdraw and abstain which makes a totalitarian regime impossible.

It is the awareness of unfulfilled desires which gives a nation the feeling that it has a mission and a destiny.

We are more prone to generalize the bad than the good. We assume that the bad is more potent and contagious.

There is sublime thieving in all giving. Someone gives us all he has and we are his.

To the excessively fearful the chief characteristic of power is its arbitrariness. Man had to gain enormously in confidence before he could conceive an all-powerful God who obeys his own laws.

A great man's greatest good luck is to die at the right time.

Wise living consists perhaps less in acquiring good habits than in acquiring as few habits as possible.

The search for happiness is one of the chief sources of unhappiness.

The feeling of being hurried is not usually the result of living a full life and having no time. It is on the contrary born of a vague fear that we are wasting our life. When we do not do the one thing we ought to do, we have no time for anything else—we are the busiest people in the world.

A heresy can spring only from a system that is in full vigor.

There is probably an element of malice in the readiness to overestimate people: we are laying up for ourselves the pleasure of later cutting them down to size.

Man was nature's mistake—she neglected to finish him—and she has never ceased paying for her mistake.

The ignorant are a reservoir of daring. It almost seems that those who have yet to discover the known are particularly equipped for dealing with the unknown. The unlearned have often rushed in where the learned feared to tread, and it is the credulous who are tempted to attempt the impossible. They know not whither they are going, and give chance a chance.

When people are free to do as they please, they usually imitate each other.

A society which gives unlimited freedom to the individual, more often than not attains a disconcerting sameness. On the other hand, where communal discipline is strict but not ruthless…originality is likely to thrive.

It almost seems that nobody can hate America as much as native Americans. America needs new immigrants to love and cherish it.

It is the individual only who is timeless. Societies, cultures, and civilizations—past and present—are often incomprehensible to outsiders, but the individual's hungers, anxieties, dreams, and preoccupations have remained unchanged through the millenia.

It would be difficult to exaggerate the degree to which we are influenced by those we influence.

People who bite the hand that feeds them usually lick the boot that kicks them.

Intolerance is the "Do Not Touch" sign on something that cannot bear touching. We do not mind having our hair ruffled, but we will not tolerate any familiarity with the toupee which covers our baldness.

It is futile to judge a kind deed by its motives. Kindness can become its own motive. We are made kind by being kind.

In a time of drastic change it is the learners who inherit the future. The learned usually find themselves equipped to live in a world that no longer exists.

The self-styled intellectual who is impotent with pen and ink hungers to write history with sword and blood.

There would be no society if living together depended upon understanding each other.

Nature is a self-made machine, more perfectly automated than any automated machine. To create something in the image of nature is to create a machine, and it was by learning the inner working of nature that man became a builder of machines.

It is remarkable by how much a pinch of malice enhances the penetrating power of an idea or an opinion. Our ears, it seems, are wonderfully attuned to sneers and evil reports about our fellow men.

There is a grandeur in the uniformity of the mass. When a fashion, a dance, a song, a slogan or a joke sweeps like wildfire from one end of the continent to the other, and a hundred million people roar with laughter, sway their bodies in unison, hum one song or break forth in anger and denunciation, there is the overpowering feeling that in this country we have come nearer the brotherhood of man than ever before.

The real antichrist is he who turns the wine of an original idea into the water of mediocrity.

A dissenting minority feels free only when it can impose its will on the majority: what it abominates most is the dissent of the majority.

It is a sign of creeping inner death when we can no longer praise the living.

A soul that is reluctant to share does not as a rule have much of its own. Miserliness is here a symptom of meagerness.

When you automate an industry you modernize it; when you automate a life you primitivize it.

It is the stretched soul that makes music, and souls are stretched by the pull of opposites—opposite bents, tastes, yearnings, loyalties. Where there is no polarity—where energies flow smoothly in one direction—there will be much doing but no music.

Naïveté in grownups is often charming; but when coupled with vanity it is indistinguishable from stupidity.

Nationalist pride, like other variants of pride, can be a substitute for self-respect.

The necessary has never been man's top priority. The passionate pursuit of the nonessential and the extravagant is one of the chief traits of human uniqueness. Unlike other forms of life, man's greatest exertions are made in the pursuit not of necessities but of superfluities.

The nineteenth century planted the words which the twentieth ripened into the atrocities of Stalin and Hitler. There is hardly an atrocity committed in the twentieth century that was not foreshadowed or even advocated by some noble man of words in the nineteenth.

To the old, the new is usually bad news.

The birth of the new constitutes a crisis, and its mastery calls for a crude and simple cast of mind—the mind of a fighter—in which the virtues of tribal cohesion and fierceness and infantile credulity and malleability are paramount. Thus every new beginning recapitulates in some degree man's first beginning.

More significant than the fact that poets write abstrusely, painters paint abstractly, and composers compose unintelligible music is that people should admire what they cannot understand; indeed, admire that which has no meaning or principle.

To spell out the obvious is often to call it in question.

The end comes when we no longer talk with ourselves. It is the end of genuine thinking and the beginning of the final loneliness.

Old age equalizes—we are aware that what is happening to us has happened to untold numbers from the beginning of time. When we are young we act as if we were the first young people in the world.

The best part of the art of living is to know how to grow old gracefully.

It still holds true that man is most uniquely human when he turns obstacles into opportunities.

Perhaps our originality manifests itself most strikingly in what we do with that which we did not originate. To discover something wholly new can be a matter of chance, of idle tinkering, or even of the chronic dissatisfaction of the untalented.

The remarkable thing is that we really love our neighbor as ourselves: we do unto others as we do unto ourselves. We hate others when we

hate ourselves. We are tolerant toward others when we tolerate ourselves. We forgive others when we forgive ourselves. We are prone to
sacrifice others when we are ready to sacrifice ourselves.

There is in most passions a shrinking away from ourselves. The passionate
pursuer has all the earmarks of a fugitive.

That which corrodes the souls of the persecuted is the monstrous inner
agreement with the prevailing prejudice against them.

The real persuaders are our appetites, our fears and above all our vanity.
The skillful propagandist stirs and coaches these internal persuaders.

It is the child in man that is the source of his uniqueness and creativeness,
and the playground is the optimal milieu for the unfolding of his
capacities and talents.

Those in possession of absolute power can not only prophesy and
make their prophecies come true, but they can also lie and make their
lies come true.

The unpredictability inherent in human affairs is due largely to the
fact that the by-products of a human process are more fateful than
the product.

Sometimes we feel the loss of a prejudice as a loss of vigor.

By all odds, earliest man, so naked to the elements and to deadly enemies, should have existed in a state of constant shock. We find him
instead the only lighthearted being in a deadly serious universe....He
alone, with childish carelessness, tinkered and played, and exerted
himself more in the pursuit of superfluities than of necessities. Yet the

tinkering and playing, and the fascination with the nonessential, were a chief source of the inventiveness which enabled man to prevail over better-equipped and more-purposeful animals.

Propaganda does not deceive people; it merely helps them to deceive themselves.

We all have private ails. The troublemakers are they who need public cures for their private ails.

We need not only a purpose in life to give meaning to our existence but also something to give meaning to our suffering. We need as much something to suffer for as something to live for.

To know a person's religion we need not listen to his profession of faith but must find his brand of intolerance.

To have a grievance is to have a purpose in life.

It not infrequently happens that those who hunger for hope give their allegiance to him who offers them a grievance.

The main effect of a real revolution is perhaps that it sweeps away those who do not know how to wish, and brings to the front men with insatiable appetites for action, power and all that the world has to offer.

We used to think that revolutions are the cause of change. Actually it is the other way around: change prepares the ground for revolution.

The savior who wants to turn men into angels is as much a hater of human nature as the totalitarian despot who wants to turn them into puppets.

Where everything is possible miracles become commonplaces, but the familiar ceases to be self-evident.

We lie loudest when we lie to ourselves.

Self-esteem and self-contempt have specific odors; they can be smelled.

No matter what our achievements might be, we think well of ourselves only in rare moments. We need people to bear witness against our inner judge, who keeps book on our shortcomings and transgressions. We need people to convince us that we are not as bad as we think we are.

Sensuality reconciles us with the human race. The misanthropy of the old is due in large part to the fading of the magic glow of desire.

A successful social technique consists perhaps in finding unobjectionable means for individual self-assertion.

Social improvement is attained more readily by a concern with the quality of results than with the purity of motives.

With some people solitariness is an escape not from others but from themselves. For they see in the eyes of others only a reflection of themselves.

A man by himself is in bad company.

The weakness of a soul is proportionate to the number of truths that must be kept from it.

Our passionate preoccupation with the sky, the stars, and a God some-where in outer space is a homing impulse. We are drawn back to where we came from.

Man is eminently a storyteller. His search for a purpose, a cause, an ideal, a mission and the like is largely a search for a plot and a pattern in the development of his life story—a story that is basically without meaning or pattern.

An empty head is not really empty; it is stuffed with rubbish. Hence the difficulty of forcing anything into an empty head.

The suspicious mind believes more than it doubts. It believes in a for-midable and ineradicable evil lurking in every person.

We are told that talent creates its own opportunities. But it sometimes seems that intense desire creates not only its own opportunities, but its own talents.

We never say so much as when we do not quite know what we want to say. We need few words when we have something to say, but all the words in all the dictionaries will not suffice when we have nothing to say and want desperately to say it.

Where there is the necessary technical skill to move mountains, there is no need for the faith that moves mountains.

There is a totalitarian regime inside every one of us. We are ruled by a ruthless politburo which sets our norms and drives us from one five-year plan to another. The autonomous individual who has to justify his existence by his own efforts is in eternal bondage to himself.

The superficiality of the American is the result of his hustling. It needs leisure to think things out; it needs leisure to mature. People in a hurry cannot think, cannot grow, nor can they decay. They are preserved in a state of perpetual puerility.

Power corrupts the few, while weakness corrupts the many....The resentment of the weak does not spring from any injustice done to them but from the sense of their inadequacy and impotence. They hate not wickedness but weakness. When it is in their power to do so, the weak destroy weakness wherever they see it.

It is a talent of the weak to persuade themselves that they suffer for something when they suffer from something; that they are showing the way when they are running away; that they see the light when they feel the heat; that they are chosen when they are shunned.

The wisdom of others remains dull till it is writ over with our own blood. We are essentially apart from the world; it bursts into our consciousness only when it sinks its teeth and nails into us.

Youth itself is a talent—a perishable talent.

David Hume

(1711-76)
Scottish philosopher, historian

Custom, then, is the great guide of human life.

Avarice, the spur of industry.

The great end of all human industry is the attainment of happiness. For this were arts invented, sciences cultivated, laws ordained, and societies modelled, by the most profound wisdom of patriots and legislators. Even the lonely savage, who lies exposed to the inclemency of the elements and the fury of wild beasts, forgets not, for a moment, this grand object of his being.

William James

(1842-1910)
U.S. psychologist, philosopher

We, the lineal representatives of the successful enactors of one scene of slaughter after another, must, whatever more pacific virtues we may also possess, still carry about with us, ready at any moment to burst into flame, the smoldering and sinister traits of character by means of which they lived through so many massacres, harming others, but themselves unharmed.

The sway of alcohol over mankind is unquestionably due to its power to stimulate the mystical faculties of human nature, usually crushed to earth by the cold facts and dry criticisms of the sober hour. Sobriety diminishes, discriminates, and says no; drunkenness expands, unites, and says yes.

If merely "feeling good" could decide, drunkenness would be the supremely valid human experience.

It is well for the world that in most of us, by the age of thirty, the character has set like plaster, and will never soften again.

Hardly ever can a youth transferred to the society of his betters unlearn the nasality and other vices of speech bred in him by the associations of his growing years. Hardly ever, indeed, no matter how much money there be in his pocket, can he ever learn to *dress* like a gentleman-born. The merchants offer their wares as eagerly to him as to the veriest "swell," but he simply *cannot* buy the right things.

Our normal waking consciousness, rational consciousness as we call it, is but one special type of consciousness, whilst all about it, parted from it by the filmiest of screens, there lie potential forms of consciousness entirely different.

The attitude of unhappiness is not only painful, it is mean and ugly. What can be more base and unworthy than the pining, puling, mumping mood, no matter by what outward ills it may have been engendered? What is more injurious to others? What less helpful as a way out of the difficulty? It but fastens and perpetuates the trouble which occasioned it, and increases the total evil of the situation. At all costs, then, we ought to reduce the sway of that mood; we ought to scout it in ourselves and others, and never show it tolerance.

The world is all the richer for having a devil in it, *so long as we keep our foot upon his neck.*

A chain is no stronger than its weakest link, and life is after all a chain.

Our esteem for facts has not neutralized in us all religiousness. It is itself almost religious. Our scientific temper is devout.

Failure, then, failure! so the world stamps us at every turn. We strew it with our blunders, our misdeeds, our lost opportunities, with all the memorials of our inadequacy to our vocation. And with what a damning emphasis does it then blot us out! No easy fine, no mere apology or formal expiation, will satisfy the world's demands, but every pound of flesh exacted is soaked with all its blood. The subtlest forms of suffering known to man are connected with the poisonous humiliations incidental to these results.

Our faith is faith in someone else's faith, and in the greatest matters this is most the case.

Individuality is founded in feeling; and the recesses of feeling, the darker, blinder strata of character, are the only places in the world in which we catch real fact in the making, and directly perceive how events happen, and how work is actually done.

Fatalism, whose solving word in all crises of behavior is "All striving is vain," will never reign supreme, for the impulse to take life strivingly is indestructible in the race. Moral creeds which speak to that impulse will be widely successful in spite of inconsistency, vagueness, and shadowy determination of expectancy. Man needs a rule for his will, and will invent one if one be not given him.

Smitten as we are with the vision of social righteousness, a God indifferent to everything but adulation, and full of partiality for his individual favorites, lacks an essential element of largeness.

Habit is thus the enormous fly-wheel of society, its most precious conservative agent. It alone is what keeps us all within the bounds of ordinance, and saves the children of fortune from the envious uprisings of the poor.

Mankind's common instinct for reality…has always held the world to be essentially a theatre for heroism. In heroism, we feel, life's supreme mystery is hidden. We tolerate no one who has no capacity whatever for it in any direction. On the other hand, no matter what a man's frailties otherwise may be, if he be willing to risk death, and still more if he suffer it heroically, in the service he has chosen, the fact consecrates him forever.

The further limits of our being plunge, it seems to me, into an altogether other dimension of existence from the sensible and merely "understandable" world. Name it the mystical region, or the supernatural region, whichever you choose. So far as our ideal impulses originate in this region (and most of them do originate in it, for we find them possessing us in a way for which we cannot articulately account), we belong to it in a more intimate sense than that in which we belong to the visible world, for we belong in the most intimate sense wherever our ideals belong.

There is no more miserable human being than one in whom nothing is habitual but indecision, and for whom the lighting of every cigar, the drinking of every cup, the time of rising and going to bed every day, and the beginning of every bit of work, are subjects of express volitional deliberation.

Knowledge about life is one thing; effective occupation of a place in life, with its dynamic currents passing through your being, is another.

We are doomed to cling to a life even while we find it unendurable.

When we of the so-called better classes are scared as men were never scared in history at material ugliness and hardship; when we put off marriage until our house can be artistic, and quake at the thought of having a child without a bank-account and doomed to manual labor, it is time for thinking men to protest against so unmanly and irreligious a state of opinion.

Metaphysics means nothing but an unusually obstinate effort to think clearly.

A little cooling down of animal excitability and instinct, a little loss of animal toughness, a little irritable weakness and descent of the pain-threshold, will bring the worm at the core of all our usual springs of delight into full view, and turn us into melancholy metaphysicians.

The prevalent fear of poverty among the educated classes is the worst moral disease from which our civilization suffers.

If the grace of God miraculously operates, it probably operates through the subliminal door.

For morality life is a war, and the service of the highest is a sort of cosmic patriotism which also calls for volunteers.

As there is no worse lie than a truth misunderstood by those who hear it, so reasonable arguments, challenges to magnanimity, and appeals to sympathy or justice, are folly when we are dealing with human crocodiles and boa-constrictors.

To be a real philosopher all that is necessary is to hate some one else's type of thinking.

I know that you, ladies and gentlemen, have a philosophy, each and all of you, and that the most interesting and important thing about you is the way in which it determines the perspective in your several worlds.

Philosophy is at once the most sublime and the most trivial of human pursuits.

We have grown literally afraid to be poor. We despise anyone who elects to be poor in order to simplify and save his inner life. If he does not join the general scramble and pant with the money-making street, we deem him spiritless and lacking in ambition.

What every genuine philosopher (every genuine man, in fact) craves most is praise—although the philosophers generally call it "recognition"!

There must be something solemn, serious, and tender about any attitude which we denominate religious. If glad, it must not grin or snicker; if sad, it must not scream or curse.

Give up the feeling of responsibility, let go your hold, resign the care of your destiny to higher powers, be genuinely indifferent as to what becomes of it all and you will find not only that you gain a perfect inward relief, but often also, in addition, the particular goods you sincerely thought you were renouncing.

Man lives for science as well as bread.

No more fiendish punishment could be devised, were such a thing physically possible, than that one should be turned loose in society and remain absolutely unnoticed.

A man has as many social selves as there are individuals who recognize him.

The moral flabbiness born of the exclusive worship of the bitch-goddess SUCCESS. That—with the squalid cash interpretation put on the word success—is our national disease.

Whatever universe a professor believes in must at any rate be a universe that lends itself to lengthy discourse. A universe definable in two sentences is something for which the professorial intellect has no use. No faith in anything of that cheap kind!

Every man who possibly can should force himself to a holiday of a full month in a year, whether he feels like taking it or not.

Immanuel Kant

(1724-1804)
German philosopher

Out of timber so crooked as that from which man is made nothing entirely straight can be carved.

Intuition and concepts constitute…the elements of all our knowledge, so that neither concepts without an intuition in some way corresponding to them, nor intuition without concepts, can yield knowledge.

All the interests of my reason, speculative as well as practical, combine in the three following questions: 1. What can I know? 2. What ought I to do? 3. What may I hope?

All thought must, directly or indirectly, by way of certain characters, relate ultimately to intuitions, and therefore, with us, to sensibility, because in no other way can an object be given to us.

Søren Kierkegaard

(1813-55)
Danish philosopher

Adversity draws men together and produces beauty and harmony in life's relationships, just as the cold of winter produces ice-flowers on the window-panes, which vanish with the warmth.

Since boredom advances and boredom is the root of all evil, no wonder, then, that the world goes backwards, that evil spreads. This can be traced back to the very beginning of the world. The gods were bored; therefore they created human beings.

I...begin with the principle that all men are bores. Surely no one will prove himself so great a bore as to contradict me in this.

I see it all perfectly; there are two possible situations—one can either do this or that. My honest opinion and my friendly advice is this: do it or do not do it—you will regret both.

Spiritual superiority only sees the individual. But alas, ordinarily we human beings are sensual and, therefore, as soon as it is a gathering, the impression changes—we see something abstract, the crowd, and we become different. But in the eyes of God, the infinite spirit, all the millions that have lived and now live do not make a crowd, He only sees each individual.

Because of its tremendous solemnity death is the light in which great passions, both good and bad, become transparent, no longer limited by outward appearances.

Nowadays not even a suicide kills himself in desperation. Before taking the step he deliberates so long and so carefully that he literally chokes with thought. It is even questionable whether he ought to be called a suicide, since it is really thought which takes his life. He does not die with deliberation but *from* deliberation.

In addition to my other numerous acquaintances, I have one more intimate confidant....My depression is the most faithful mistress I have known—no wonder, then, that I return the love.

It belongs to the imperfection of everything human that man can only attain his desire by passing through its opposite.

Doubt is thought's despair; despair is personality's doubt....Doubt and despair...belong to completely different spheres; different sides of the soul are set in motion....Despair is an expression of the total personality, doubt only of thought.

At the bottom of enmity between strangers lies indifference.

Listen to the cry of a woman in labor at the hour of giving birth—look at the dying man's struggle at his last extremity, and then tell me whether something that begins and ends thus could be intended for enjoyment.

I do not care for anything. I do not care to ride, for the exercise is too violent. I do not care to walk, walking is too strenuous. I do not care to lie down, for I should either have to remain lying, and I do not care to do that, or I should have to get up again, and I do not care to do that either. *Summa summarum*: I do not care at all.

Faith is the highest passion in a human being. Many in every generation may not come that far, but none comes further.

How absurd men are! They never use the liberties they have, they demand those they do not have. They have freedom of thought, they demand freedom of speech.

I feel as if I were a piece in a game of chess, when my opponent says of it: That piece cannot be moved.

Concepts, like individuals, have their histories and are just as incapable of withstanding the ravages of time as are individuals. But in and through all this they retain a kind of homesickness for the scenes of their childhood.

Not just in commerce but in the world of ideas too our age is putting on a veritable clearance sale. Everything can be had so dirt cheap that one begins to wander whether in the end anyone will want to make a bid.

The present generation, wearied by its chimerical efforts, relapses into complete indolence. Its condition is that of a man who has only fallen asleep towards morning: first of all come great dreams, then a feeling of laziness, and finally a witty or clever excuse for remaining in bed.

There are, as is known, insects that die in the moment of fertilization. So it is with all joy: life's highest, most splendid moment of enjoyment is accompanied by death.

It is quite true what Philosophy says: that Life must be understood backwards. But that makes one forget the other saying: that it must be lived—forwards. The more one ponders this, the more it comes to mean that life in the temporal existence never becomes quite intelligible, precisely because at no moment can I find complete quiet to take the backward-looking position.

This is what is sad when one contemplates human life, that so many live out their lives in quiet lostness…they live, as it were, away from themselves and vanish like shadows. Their immortal souls are blown away, and they are not disquieted by the question of its immortality, because they are already disintegrated before they die.

Marriage brings one into fatal connection with custom and tradition, and traditions and customs are like the wind and weather, altogether incalculable.

The difference between a man who faces death for the sake of an idea and an imitator who goes in search of martyrdom is that whilst the former expresses his idea most fully in death it is the strange feeling of bitterness which comes from failure that the latter really enjoys; the former rejoices in his victory, the latter in his suffering.

The more a man can forget, the greater the number of metamorphoses which his life can undergo, the more he can remember the more divine his life becomes.

Truth always rests with the minority, and the minority is always stronger than the majority, because the minority is generally formed by those who really have an opinion, while the strength of a majority is illusory, formed by the gangs who have no opinion—and who, therefore, in the next instant (when it is evident that the minority is the stronger) assume its opinion…while Truth again reverts to a new minority.

Just as in earthly life lovers long for the moment when they are able to breathe forth their love for each other, to let their souls blend in a soft whisper, so the mystic longs for the moment when in prayer he can, as it were, creep into God.

Old age realizes the dreams of youth: look at Dean Swift; in his youth he built an asylum for the insane, in his old age he was himself an inmate.

The paradox is really the *pathos* of intellectual life and just as only great souls are exposed to passions it is only the great thinker who is exposed to what I call paradoxes, which are nothing else than grandiose thoughts in embryo.

Personality is only ripe when a man has made the truth his own.

Philosophy always requires something more, requires the eternal, the true, in contrast to which even the fullest existence as such is but a happy moment.

Most men pursue pleasure with such breathless haste that they hurry past it.

What is a poet? An unhappy person who conceals profound anguish in his heart but whose lips are so formed that as sighs and cries pass over them they sound like beautiful music.

At one time my only wish was to be a police official. It seemed to me to be an occupation for my sleepless intriguing mind. I had the idea that there, among criminals, were people to fight: clever, vigorous, crafty fellows. Later I realized that it was good that I did not become one, for most police cases involve misery and wretchedness—not crimes and scandals.

If I were to wish for anything, I should not wish for wealth and power, but for the passionate sense of the potential, for the eye which, ever young and ardent, sees the possible. Pleasure disappoints, possibility never. And what wine is so sparkling, what so fragrant, what so intoxicating, as possibility!

Father in Heaven! When the thought of thee wakes in our hearts let it not awaken like a frightened bird that flies about in dismay, but like a child waking from its sleep with a heavenly smile.

Destroy your primitivity, and you will most probably get along well in the world, maybe achieve great success—but Eternity will reject you. Follow up your primitivity, and you will be shipwrecked in temporality, but accepted by Eternity.

The most terrible fight is not when there is one opinion against another, the most terrible is when two men say the same thing—and fight about the interpretation, and this interpretation involves a difference of quality.

God creates out of *nothing*, wonderful, you say: yes, to be sure, but he does what is still more wonderful: he makes saints out of sinners.

Since my earliest childhood a barb of sorrow has lodged in my heart. As long as it stays I am ironic—if it is pulled out I shall die.

It requires courage not to surrender oneself to the ingenious or compassionate counsels of despair that would induce a man to eliminate himself from the ranks of the living; but it does not follow from this that every huckster who is fattened and nourished in self-confidence has more courage than the man who yielded to despair.

How ironical that it is by means of speech that man can degrade himself below the level of dumb creation—for a chatterbox is truly of a lower category than a dumb creature.

People commonly travel the world over to see rivers and mountains, new stars, garish birds, freak fish, grotesque breeds of human; they fall into an animal stupor that gapes at existence and they think they have seen something.

In order to swim one takes off all one's clothes—in order to aspire to the truth one must undress in a far more inward sense, divest oneself of all one's inward clothes, of thoughts, conceptions, selfishness etc., before one is sufficiently naked.

The truth is a snare: you cannot have it, without being caught. You cannot have the truth in such a way that you catch it, but only in such a way that it catches you.

It is the duty of the human understanding to understand that there are things which it cannot understand, and what those things are. Human understanding has vulgarly occupied itself with nothing but under-standing, but if it would only take the trouble to understand itself at the same time it would simply have to posit the paradox.

Georg Christoph Lichtenberg

(1742-99)
German physicist, philosopher

Man can acquire accomplishments or he can become an animal, whichever he wants. God makes the animals, man makes himself.

The sure conviction that we could if we wanted to is the reason so many good minds are idle.

It is in the gift for employing all the vicissitudes of life to one's own advantage and to that of one's craft that a large part of genius consists.

Affectation is a very good word when someone does not wish to confess to what he would none the less like to believe of himself.

One is rarely an impulsive innovator after the age of sixty, but one can still be a very fine orderly and inventive thinker. One rarely procreates children at that age, but one is all the more skilled at educating those who have already been procreated, and education is procreation of another kind.

To receive applause for works which do not demand all our powers hinders our advance towards a perfecting of our spirit. It usually means that thereafter we stand still.

Actual aristocracy cannot be abolished by any law: all the law can do is decree how it is to be imparted and who is to acquire it.

Astronomy is perhaps the science whose discoveries owe least to chance, in which human understanding appears in its whole magnitude, and through which man can best learn how small he is.

Much reading has brought upon us a learned barbarism.

First we have to *believe*, and then we believe.

With most people disbelief in a thing is founded on a blind belief in some other thing.

Many things about our bodies would not seem to us so filthy and obscene if we did not have the idea of nobility in our heads.

A book is a mirror: if an ape looks into it an apostle is hardly likely to look out.

Do we write books so that they shall merely be read? Don't we also write them for employment in the household? For one that is read from start to finish, thousands are leafed through, other thousands lie motionless, others are jammed against mouseholes, thrown at rats, others are stood on, sat on, drummed on, have gingerbread baked on them or are used to light pipes.

There were honest people long before there were Christians and there are, God be praised, still honest people where there are no Christians. It could therefore easily be possible that people are Christians because true Christianity corresponds to what they would have been even if Christianity did not exist.

Man loves company, even if it is only that of a smouldering candle.

To be content with life—or to live merrily, rather—all that is required is that we bestow on all things only a fleeting, superficial glance; the more thoughtful we become the more earnest we grow.

The fly that does not want to be swatted is safest if it sits on the fly-swat.

Once the good man was dead, one wore his hat and another his sword as he had worn them, a third had himself barbered as he had, a fourth walked as he did, but the honest man that he was—nobody any longer wanted to be that.

If we make a couple of discoveries here and there we need not believe things will go on like this for ever....Just as we hit water when we dig in the earth, so we discover the incomprehensible sooner or later.

Doubt must be no more than vigilance, otherwise it can become dangerous.

To err is *human* also in so far as animals seldom or never err, or at least only the cleverest of them do so.

What is the good of drawing conclusions from experience? I don't deny we sometimes draw the right conclusions, but don't we just as often draw the wrong ones?

We can see nothing whatever of the soul unless it is visible in the expression of the countenance; one might call the faces at a large assembly of people a history of the human soul written in a kind of Chinese ideograms.

Once we know our weaknesses they cease to do us any harm.

Even truth needs to be clad in new garments if it is to appeal to a new age.

He who says he hates every kind of flattery, and says it in earnest, certainly does not yet know every kind of flattery.

Food probably has a very great influence on the condition of men. Wine exercises a more visible influence, food does it more slowly but perhaps just as surely. Who knows if a well-prepared soup was not responsible for the pneumatic pump or a poor one for a war?

A clever child brought up with a foolish one can itself become foolish. Man is so perfectable and corruptible he can become a fool through good sense.

Man is a masterpiece of creation if for no other reason than that, all the weight of evidence for determinism notwithstanding, he believes he has free will.

What most clearly characterizes true freedom and its true employment is its misemployment.

What I do not like about our definitions of genius is that there is in them nothing of the day of judgment, nothing of resounding through eternity and nothing of the footsteps of the Almighty.

There is no more important rule of conduct in the world than this: attach yourself as much as you can to people who are abler than you and yet not so very different that you cannot understand them.

There are people who possess not so much genius as a certain talent for perceiving the desires of the century, or even of the decade, before it has done so itself.

The Greeks possessed a knowledge of human nature we seem hardly able to attain to without passing through the strengthening hibernation of a new barbarism.

One might call habit a moral friction: something that prevents the mind from gliding over things but connects it with them and makes it hard for it to free itself from them.

Of all the inventions of man I doubt whether any was more easily accomplished than that of a Heaven.

What is called an acute knowledge of human nature is mostly nothing but the observer's own weaknesses reflected back from others.

That man is the noblest creature may also be inferred from the fact that no other creature has yet contested this claim.

If you are going to build something in the air it is always better to build castles than houses of cards.

Ideas too are a life and a world.

I believe that man is in the last resort so free a being that his right *to be* what he believes himself to be cannot be contested.

Sickness is mankind's greatest defect.

To do the opposite of something is also a form of imitation, namely an imitation of its opposite.

Man is always partial and is quite right to be. Even impartiality is partial.

Nothing can contribute more to peace of soul than the lack of any opinion whatever.

The greatest events occur without intention playing any part in them; chance makes good mistakes and undoes the most carefully planned undertaking. The world's greatest events are not produced, they happen.

If all else fails, the character of a man can be recognized by nothing so surely as by a jest which he takes badly.

The journalists have constructed for themselves a little wooden chapel, which they also call the Temple of Fame, in which they put up and take down portraits all day long and make such a hammering you can't hear yourself speak.

Erudition can produce foliage without bearing fruit.

Before we blame we should first see whether we cannot excuse.

It is almost everywhere the case that soon after it is begotten the greater part of human wisdom is laid to rest in *repositories*.

With a pen in my hand I have successfully stormed bulwarks from which others armed with sword and excommunication have been repulsed.

A good metaphor is something even the police should keep an eye on.

It is no great art to say something briefly when, like Tacitus, one has something to say; when one has nothing to say, however, and none the less writes a whole book and makes truth…into a liar—that I call an achievement.

So-called professional mathematicians have, in their reliance on the relative incapacity of the rest of mankind, acquired for themselves a reputation for profundity very similar to the reputation for sanctity possessed by theologians.

Much can be inferred about a man from his mistress: in her one beholds his weaknesses and his dreams.

Every man has his moral backside which he refrains from showing unless he has to and keeps covered as long as possible with the trousers of decorum.

He who is enamored of himself will at least have the advantage of being inconvenienced by few rivals.

We cannot remember too often that when we observe nature, and especially the ordering of nature, it is always ourselves alone we are observing.

If people should ever start to do only what is necessary millions would die of hunger.

The American who first discovered Columbus made a bad discovery.

Be wary of passing the judgment: *obscure*. To find something obscure poses no difficulty: elephants and poodles find many things obscure.

Nothing makes one old so quickly as the ever-present thought that one is growing older.

I have remarked very clearly that I am often of one opinion when I am lying down and of another when I am standing up.

We accumulate our opinions at an age when our understanding is at its weakest.

A handful of soldiers is always better than a mouthful of arguments.

We are obliged to regard many of our original minds as crazy at least until we have become as clever as they are.

I am convinced we do not only love ourselves in others but hate ourselves in others too.

Man is to be found in reason, God in the passions.

We often have need of a profound philosophy to restore to our feelings their original state of innocence, to find *our* way out of the rubble of things alien to us, to begin to feel for *ourselves* and to speak *ourselves*, and I might almost say to exist ourselves.

Even if my philosophy does not extend to discovering anything new it does nevertheless.

It is hardly to be believed how spiritual reflections when mixed with a little physics can hold people's attention and give them a livelier idea of God than do the often ill-applied examples of his wrath.

Prejudices are so to speak the mechanical instincts of men: through their prejudices they do without any effort many things they would find too difficult to think through to the point of resolving to do them.

The most dangerous untruths are truths slightly distorted.

With prophecies the commentator is often a more important man than the prophet.

There exists a species of transcendental ventriloquism by means of which men can be made to believe that something said on earth comes from Heaven.

We say that someone occupies an official position, whereas it is the official position that occupies him.

If another Messiah was born he could hardly do so much good as the printing-press.

There are very many people who read simply to prevent themselves from thinking.

It is said that truth comes from the mouths of fools and children: I wish every good mind which feels an inclination for satire would reflect that the finest satirist always has something of both in him.

One cannot demand of a scholar that he show himself a scholar every-where in society, but the whole tenor of his behavior must none the less betray the thinker, he must always be instructive, his way of judging a thing must even in the smallest matters be such that people can see what it will amount to when, quietly and self-collected, he puts this power to scholarly use.

People often become scholars for the same reason they become soldiers: simply because they are unfit for any other station. Their right hand has to earn them a livelihood; one might say they lie down like bears in winter and seek sustenance from their paws.

The most heated defenders of a science, who cannot endure the slightest sneer at it, are commonly those who have not made very much progress in it and are secretly aware of this defect.

There is no greater impediment to progress in the sciences than the desire to see it take place too quickly.

The "second sight" possessed by the Highlanders in Scotland is actually a foreknowledge of future events. I believe they possess this gift because they don't wear trousers. That is also why in all countries women are more prone to utter prophecies.

He was always smoothing and polishing himself, and in the end he became blunt before he was sharp.

To grow wiser means to learn to know better and better the faults to which this instrument with which we feel and judge can be subject.

The noble simplicity in the works of nature only too often originates in the noble shortsightedness of him who observes it.

Cautiousness in judgment is nowadays to be recommended to each and every one: if we gained only one incontestable truth every ten years from each of our philosophical writers the harvest we reaped would be sufficient.

When an acquaintance goes by I often step back from my window, not so much to spare him the effort of acknowledging me as to spare myself the embarrassment of seeing that he has not done so.

There are people who believe everything is sane and sensible that is done with a solemn face.

Just as the performance of the vilest and most wicked deeds requires spirit and talent, so even the greatest demand a certain insensitivity which under other circumstances we would call stupidity.

The most perfect ape cannot draw an ape; only man can do that; but, likewise, only man regards the ability to do this as a sign of superiority.

The great rule: If the little bit you have is nothing special in itself, at least find a way of saying it that is a little bit special.

Good taste is either that which agrees with my taste or that which subjects itself to the rule of reason. From this we can see how useful it is to employ reason in seeking out the laws of taste.

A schoolteacher or professor cannot educate individuals, he educates only species.

The most successful tempters and thus the most dangerous are the deluded deluders.

Delight at having understood a very abstract and obscure system leads most people to believe in the truth of what it demonstrates.

As the few adepts in such things well know, universal morality is to be found in little everyday penny-events just as much as in great ones. There is so much goodness and ingenuity in a raindrop that an apothecary wouldn't let it go for less than half-a-crown.

The human tendency to regard little things as important has produced very many great things.

Rational free spirits are the light brigade who go on ahead and reconnoitre the ground which the heavy brigade of the orthodox will eventually occupy.

Virtue by premeditation isn't worth much.

We have no words for speaking of wisdom to the stupid. He who understands the wise is wise already.

As I take up my pen I feel myself so full, so equal to my subject, and see my book so clearly before me in embryo, I would almost like to try to say it all in a single word.

John Locke

(1632-1704)
English philosopher

The only fence against the world is a thorough knowledge of it.

Freedom of men under government is to have a standing rule to live by, common to every one of that society, and made by the legislative power vested in it; a liberty to follow my own will in all things, when the rule prescribes not, and not to be subject to the inconstant, unknown, arbitrary will of another man.

A sound mind in a sound body, is a short, but full description of a happy state in this World: he that has these two, has little more to wish for; and he that wants either of them, will be little the better for anything else.

New opinions are always suspected, and usually opposed, without any other reason but because they are not already common.

Lucretius

(c.99 B.C.-c.55 B.C.)
Roman poet, philosopher

Pleasant it is, when over a great sea the winds trouble the waters, to gaze from shore upon another's great tribulation; not because any man's troubles are a delectable joy, but because to perceive you are free of them yourself is pleasant.

From the very fountain of enchantment there arises a taste of bitterness to spread anguish amongst the flowers.

Niccolò Machiavelli

(1469-1527)
Italian political philosopher, statesman

It should be noted that when he seizes a state the new ruler ought to determine all the injuries that he will need to inflict. He should inflict them once and for all, and not have to renew them every day. Whoever acts otherwise, either through timidity or bad advice, is always forced to have the knife ready in his hand....Violence should be inflicted once and for all; people will then forget what it tastes like and so be less resentful.

Benefits should be conferred gradually; and in that way they will taste better.

The wish to acquire more is admittedly a very natural and common thing; and when men succeed in this they are always praised rather than condemned. But when they lack the ability to do so and yet want to acquire more at all costs, they deserve condemnation for their mistakes.

Many have dreamed up republics and principalities that have never in truth been known to exist; the gulf between how one should live and how one does live is so wide that that a man who neglects what is actually done for what should be done learns the way to self-destruction rather than self-preservation.

Men nearly always follow the tracks made by others and proceed in their affairs by imitation, even though they cannot entirely keep to the tracks of others or emulate the prowess of their models. So a prudent man should always follow in the footsteps of great men and imitate those who have been outstanding. If his own prowess fails to compare with theirs, at least it has an air of greatness about it. He should behave like those archers who, if they are skillful, when the target seems too distant, know the capabilities of their bow and aim a good deal higher than their objective, not in order to shoot so high but so that by aiming high they can reach the target.

Men sooner forget the death of their father than the loss of their patrimony.

There are three kinds of intelligence: one kind understands things for itself, the other appreciates what others can understand, the third understands neither for itself nor through others. This first kind is excellent, the second good, and the third kind useless.

States that rise quickly, just as all the other things of nature that are born and grow rapidly, cannot have roots and ramifications; the first bad weather kills them.

A prince never lacks legitimate reasons to break his promise.

Since it is difficult to join them together, it is safer to be feared than to be loved when one of the two must be lacking.

A prince must be prudent enough to know how to escape the bad reputation of those vices that would lose the state for him, and must protect himself from those that will not lose it for him, if this is possible;

but if he cannot, he need not concern himself unduly if he ignores these less serious vices.

The fact is that a man who wants to act virtuously in every way necessarily comes to grief among so many who are not virtuous.

There is no avoiding war; it can only be postponed to the advantage of others.

For among other evils caused by being disarmed, it renders you contemptible; which is one of those disgraceful things which a prince must guard against.

Moses Maimonides

(1135-1204)
Egyptian physician, philosopher

There are eight rungs in charity. The highest is when you help a man to help himself.

Joseph de Maistre

(1753-1821)
French diplomat, philosopher

All grandeur, all power, all subordination to authority rests on the executioner: he is the horror and the bond of human association. Remove this incomprehensible agent from the world and at that very moment order gives way to chaos, thrones topple and society disappears.

A constitution that is made for all nations is made for none.

Man is insatiable for power; he is infantile in his desires and, always discontented with what he has, loves only what he has not. People complain of the despotism of princes; they ought to complain of the despotism of man.

We are tainted by modern philosophy which has taught us that *all is good*, whereas evil has polluted everything and in a very real sense *all is evil*, since nothing is in its proper place.

If there was no moral evil upon earth, there would be no physical evil.

In the works of man, everything is as poor as its author; vision is confined, means are limited, scope is restricted, movements are labored, and results are humdrum.

We are all bound to the throne of the Supreme Being by a flexible chain which restrains without enslaving us. The most wonderful aspect of the universal scheme of things is the action of free beings under divine guidance.

It can even come about that a created will cancels out, not perhaps the *exertion*, but the result of divine action; for in this sense, God himself has told us that *God* wishes things which do not happen because man does not wish them! Thus the rights of men are immense, and his greatest misfortune is to be unaware of them.

Man in general, if reduced to himself, is too wicked to be free.

Without doubt God is the universal moving force, but each being is moved according to the nature that God has given it....He directs angels, man, animals, brute matter, in sum all created things, but each *according to its nature*, and man having been created free, he is freely led. This rule is truly *the eternal law* and in it we must believe.

Every country has the government it deserves.

There is no instant of time when one creature is not being devoured by another. Over all these numerous races of animals man is placed, and his destructive hand spares nothing that lives. He kills to obtain food and he kills to clothe himself; he kills to adorn himself; he kills in order to attack and he kills to defend himself; he kills to instruct himself and he kills to amuse himself; he kills to kill. Proud and terrible king, he wants everything and nothing resists him.

Nothing is necessary except God, and nothing is less necessary than pain.

False opinions are like false money, struck first of all by guilty men and thereafter circulated by honest people who perpetuate the crime without knowing what they are doing.

All pain is a punishment, and every punishment is inflicted for love as much as for justice.

There is no philosophy without the art of ignoring objections.

It is one of man's curious idiosyncrasies to create difficulties for the pleasure of resolving them.

Wherever an altar is found, there civilization exists.

There is no man who desires as passionately as a Russian. If we could imprison a Russian desire beneath a fortress, that fortress would explode.

The whole earth, perpetually steeped in blood, is nothing but an immense altar on which every living thing must be sacrificed without end, without restraint, without respite until the consummation of the world, the extinction of evil, the death of death.

Man is so muddled, so dependent on the things immediately before his eyes, that every day even the most submissive believer can be seen to risk the torments of the afterlife for the smallest pleasure.

In the whole vast dome of living nature there reigns an open violence, a kind of prescriptive fury which arms all the creatures to their common doom: as soon as you leave the inanimate kingdom you find the decree of violent death inscribed on the very frontiers of life.

War is thus divine in itself, since it is a law of the world. War is divine through its consequences of a supernatural nature which are as much general as particular....War is divine in the mysterious glory that surrounds it and in the no less inexplicable attraction that draws us to it....War is divine by the manner in which it breaks out.

Herbert Marcuse

(1898-1979)
U.S. political philosopher

Self-determination, the autonomy of the individual, asserts itself in the right to race his automobile, to handle his power tools, to buy a gun, to communicate to mass audiences his opinion, no matter how ignorant, how aggressive, it may be.

The so-called consumer society and the politics of corporate capitalism have created a second nature of man which ties him libidinally and aggressively to the commodity form. The need for possessing, consuming, handling and constantly renewing the gadgets, devices, instruments, engines, offered to and imposed upon the people, for using these wares even at the danger of one's own destruction, has become a "biological" need.

If mass communications blend together harmoniously, and often unnoticeably, art, politics, religion, and philosophy with commercials, they bring these realms of culture to their common denominator—the commodity form. The music of the soul is also the music of salesmanship. Exchange value, not truth value, counts.

If the worker and his boss enjoy the same television program and visit the same resort places, if the typist is as attractively made up as the daughter of her employer, if the Negro owns a Cadillac, if they all read the same newspaper, then this assimilation indicates not the disappearance of classes, but the extent to which the needs and satisfactions that serve the preservation of the Establishment are shared by the underlying population.

Freedom of enterprise was from the beginning not altogether a blessing. As the liberty to work or to starve, it spelled toil, insecurity, and fear for the vast majority of the population. If the individual were no longer compelled to prove himself on the market, as a free economic subject, the disappearance of this freedom would be one of the greatest achievements of civilization.

The people recognize themselves in their commodities; they find their soul in their automobile, hi-fi set, split-level home, kitchen equipment.

Obscenity is a moral concept in the verbal arsenal of the Establishment, which abuses the term by applying it, not to expressions of its own morality but to those of another.

The web of domination has become the web of Reason itself, and this society is fatally entangled in it.

Jacques Maritain

(1882-1973)
French philosopher

We don't love qualities, we love persons; sometimes by reason of their defects as well as of their qualities.

Gratitude is the most exquisite form of courtesy.

I don't see America as a mainland, but as a sea, a big ocean. Sometimes a storm arises, a formidable current develops, and it seems it will engulf everything. Wait a moment, another current will appear and bring the first one to naught.

Karl Marx

(1818-83)
German political theorist, social philosopher

The bourgeoisie of the whole world, which looks complacently upon the wholesale massacre after the battle, is convulsed by horror at the desecration of brick and mortar.

The bourgeoisie…has been the first to show what man's activity can bring about. It has accomplished wonders far surpassing Egyptian pyramids, Roman aqueducts and Gothic cathedrals.…The bourgeoisie…draws all, even the most barbarian nations into civilization.…It has created enormous cities…and has thus rescued a considerable part of the population from the idiocy of rural life.…The bourgeoisie, during its rule of scarce one hundred years, has created more massive and more colossal productive forces than have all preceding generations together.

Capital is money, capital is commodities.…By virtue of it being value, it has acquired the occult ability to add value to itself. It brings forth living offspring, or, at the least, lays golden eggs.

Capital is dead labor, which, vampire-like, lives only by sucking living labor, and lives the more, the more labor it sucks.

The history of all hitherto existing society is the history of class struggles.

In a higher phase of communist society…only then can the narrow horizon of bourgeois right be fully left behind and society inscribe on its banners: from each according to his ability, to each according to his needs.

The theory of the Communists may be summed up in the single sentence: Abolition of private property.

The ideas of the ruling class are in every epoch the ruling ideas, i.e., the class which is the ruling *material* force of society, is at the same time its ruling *intellectual* force.

The development of civilization and industry in general has always shown itself so active in the destruction of forests that everything that has been done for their conservation and production is completely insignificant in comparison.

Greek philosophy seems to have met with something with which a good tragedy is not supposed to meet, namely, a dull ending.

History repeats itself, first as tragedy, second as farce.

Men make their own history, but they do not make it just as they please; they do not make it under circumstances chosen by themselves, but under circumstances directly found, given and transmitted from the past. The tradition of all the dead generations weighs like a nightmare on the brain of the living.

We know only a single science, the science of history. One can look at history from two sides and divide it into the history of nature and the history of men. However, the two sides are not to be divided off; as long as men exist the history of nature and the history of men are mutually conditioned.

History does nothing; it does not possess immense riches, it does not fight battles. It is men, real, living, who do all this....It is not "history" which uses men as a means of achieving—as if it were an individual person—its own ends. History is nothing but the activity of men in pursuit of their ends.

In bourgeois society capital is independent and has individuality, while the living person is dependent and has no individuality.

Colonial system, public debts, heavy taxes, protection, commercial wars, etc., these offshoots of the period of manufacture swell to gigantic proportions during the period of infancy of large-scale industry.

The birth of the latter is celebrated by a vast, Herod-like slaughter of the innocents.

On a level plain, simple mounds look like hills; and the insipid flatness of our present bourgeoisie is to be measured by the altitude of its "great intellects."

The country that is more developed industrially only shows, to the less developed, the image of its own future.

Landlords, like all other men, love to reap where they never sowed.

The writer may very well serve a movement of history as its mouthpiece, but he cannot of course create it.

Machines were, it may be said, the weapon employed by the capitalists to quell the revolt of specialized labor.

What I did that was new was to prove: (1) that the existence of classes is only bound up with particular, historic phases in the development of production; (2) that the class struggle necessarily leads to the dictatorship of the proletariat; (3) that this dictatorship itself only constitutes the transition to the abolition of all classes and to a classless society.

All I know is I'm not a Marxist.

While the miser is merely a capitalist gone mad, the capitalist is a rational miser.

All social rules and all relations between individuals are eroded by a cash economy, avarice drags Pluto himself out of the bowels of the earth.

It is absolutely impossible to transcend the laws of nature. What can change in historically different circumstances is only the form in which these laws expose themselves.

The philosophers have only *interpreted* the world in various ways; the point, however, is to *change* it.

Philosophy stands in the same relation to the study of the actual world as masturbation to sexual love.

As in private life one differentiates between what a man thinks and says of himself and what he really is and does, so in historical struggles one must still more distinguish the language and the imaginary aspirations of parties from their real organism and their real interests, their conception of themselves from their reality.

The human being is in the most literal sense a political animal, not merely a gregarious animal, but an animal which can individuate itself only in the midst of society.

In the domain of Political Economy, free scientific inquiry meets not merely the same enemies as in all other domains. The peculiar nature of the material it deals with, summons as foes into the field of battle the most violent, mean and malignant passions of the human breast, the Furies of private interest.

Constant revolutionizing of production...distinguish the bourgeois epoch from all earlier ones. All fixed, fast-frozen relations, with their train of ancient and venerable prejudices are swept away, all new-formed ones become antiquated before they can ossify. All that is solid melts into air, all that is holy is profaned, and man is at last compelled to face with sober senses, his real conditions of life, and his relations with his kind.

Mankind always sets itself only such tasks as it can solve; since, looking at the matter more closely, we will always find that the task itself arises only when the material conditions necessary for its solution already exist or are at least in the process of formation.

Religion is the sigh of the oppressed creature, the heart of a heartless world, and the soul of soulless conditions. It is the *opium* of the people.

In every revolution there intrude, at the side of its true agents, men of a different stamp; some of them survivors of and devotees to past revolutions, without insight into the present movement, but preserving popular influence by their known honesty and courage, or by the sheer force of tradition; others mere brawlers, who, by dint of repeating year after year the same set of stereotyped declamations against the government of the day, have sneaked into the reputation of revolutionists of the first water....They are an unavoidable evil: with time they are shaken off.

A commodity appears at first sight an extremely obvious, trivial thing. But its analysis brings out that it is a very strange thing, abounding in metaphysical subtleties and theological niceties.

Natural science will in time incorporate into itself the science of man, just as the science of man will incorporate into itself natural science: there will be one science.

The product of mental labor—science—always stands far below its value, because the labor-time necessary to reproduce it has no relation at all to the labor-time required for its original production.

Society does not consist of individuals but expresses the sum of inter-relations, the relations within which these individuals stand.

Civil servants and priests, soldiers and ballet-dancers, schoolmasters and police constables, Greek museums and Gothic steeples, civil list and services list—the common seed within which all these fabulous beings slumber in embryo is taxation.

We should not say that one man's hour is worth another man's hour, but rather that one man during an hour is worth just as much as another man during an hour. Time is everything, man is nothing: he is at the most time's carcass.

The production of too many useful things results in too many useless people.

In communist society, where nobody has one exclusive sphere of activity but each can become accomplished in any branch he wishes, society regulates the general production and thus makes it possible for me to do one thing today and another tomorrow, to hunt in the morning, fish in the afternoon, rear cattle in the evening, criticize after dinner, just as I have a mind, without ever becoming hunter, fisherman, shepherd or critic.

John Stuart Mill

(1806-73)
English philosopher, economist

As for charity, it is a matter in which the immediate effect on the persons directly concerned, and the ultimate consequence to the general good, are apt to be at complete war with one another.

Though the practice of chivalry fell even more sadly short of its theoretic standard than practice generally falls below theory, it remains one of the most precious monuments of the moral history of our race, as a remarkable instance of a concerted and organized attempt by a most disorganized and distracted society, to raise up and carry into practice a moral ideal greatly in advance of its social condition and institutions; so much so as to have been completely frustrated in the main object, yet never entirely inefficacious, and which has left a most sensible, and for the most part a highly valuable impress on the ideas and feelings of all subsequent times.

That so few now dare to be eccentric, marks the chief danger of the time.

The despotism of custom is everywhere the standing hindrance to human advancement.

If all mankind minus one, were of one opinion, and only one person were of the contrary opinion, mankind would be no more justified in silencing that one person, than he, if he had the power, would be justified in silencing mankind.

Ask yourself whether you are happy, and you cease to be so.

As long as justice and injustice have not terminated their ever renewing fight for ascendancy in the affairs of mankind, human beings must be willing, when need is, to do battle for the one against the other.

A man who has nothing which he cares about more than he does about his personal safety is a miserable creature who has no chance of being free, unless made and kept so by the existing of better men than himself.

All that makes existence valuable to any one depends on the enforcement of restraints upon the actions of other people.

The only power deserving the name is that of masses, and of governments while they make themselves the organ of the tendencies and instincts of masses.

The general tendency of things throughout the world is to render mediocrity the ascendant power among mankind.

The peculiar evil of silencing the expression of an opinion is, that it is robbing the human race; posterity as well as the existing generation; those who dissent from the opinion, still more than those who hold it. If the opinion is right, they are deprived of the opportunity of exchanging error for truth: if wrong, they lose, what is almost as great a benefit, the clearer perception and livelier impression of truth, produced by its collision with error.

A party of order or stability, and a party of progress or reform, are both necessary elements of a healthy state of political life.

We can never be sure that the opinion we are endeavouring to stifle is a false opinion; and even if we were sure, stifling it would be an evil still.

The worth of a State, in the long run, is the worth of the individuals composing it....A State which dwarfs its men, in order that they may be more docile instruments in its hands even for beneficial purposes—will find that with small men no great thing can really be accomplished.

War is an ugly thing, but not the ugliest of things: the decayed and degraded state of moral and patriotic feeling which thinks nothing *worth* a war, is worse....A war to protect other human beings against tyrannical injustice; a war to give victory to their own ideas of right and good, and which is their own war, carried on for an honest purpose by their own free choice—is often the means of their regeneration.

Michel de Montaigne

(1533-92)
French essayist, philosopher

By some might be said of me that here I have but gathered a nosegay of strange flowers, and have put nothing of mine unto it but the thread to bind them.

No profession or occupation is more pleasing than the military; a profession or exercise both noble in execution (for the strongest, most generous and proudest of all virtues is true valour) and noble in its cause. No utility either more just or universal than the protection of the repose or defence of the greatness of one's country. The company and daily conversation of so many noble, young and active men cannot but be well-pleasing to you.

How many things served us but yesterday as articles of faith, which today we deem but fables?

Every abridgment of a good book is a fool abridged.

Let us not be ashamed to speak what we shame not to think.

For truly it is to be noted, that children's plays are not sports, and should be deemed as their most serious actions.

There is no pleasure to me without communication: there is not so much as a sprightly thought comes into my mind that it does not grieve me to have produced alone, and that I have no one to tell it to.

The worst of my actions or conditions seem not so ugly unto me as I find it both ugly and base not to dare to avouch for them.

There is not much less vexation in the government of a private family than in the managing of an entire state.

If a man urge me to tell wherefore I loved him, I feel it cannot be expressed but by answering: Because it was he, because it was myself.

To honor him whom we have made is far from honoring him that hath made us.

What harm cause not those huge draughts or pictures which wanton youth with chalk or coals draw in each passage, wall or stairs of our great houses, whence a cruel contempt of our natural store is bred in them?

I love those historians that are either very simple or most excellent....Such as are between both (which is the most common fashion), it is they that spoil all; they will needs chew our meat for us and take upon them a law to judge, and by consequence to square and incline the story according to their fantasy.

My art and profession is to live.

It is much more easy to accuse the one sex than to excuse the other.

Nature should have been pleased to have made this age miserable, without making it also ridiculous.

Scratching is one of nature's sweetest gratifications, and the one nearest at hand.

The worthiest man to be known, and for a pattern to be presented to the world, he is the man of whom we have most certain knowledge. He hath been declared and enlightened by the most clear-seeing men that ever were; the testimonies we have of him are in faithfulness and sufficiency most admirable.

We endeavour more that men should speak of us, than how and what they speak, and it sufficeth us that our name run in men's mouths, in what manner soever. It seemeth that to be known is in some sort to have life and continuance in other men's keeping.

The same reason that makes us chide and brawl and fall out with any of our neighbours, causeth a war to follow between Princes.

Oh senseless man, who cannot possibly make a worm, and yet will make Gods by dozens.

One may disavow and disclaim vices that surprise us, and whereto our passions transport us; but those which by long habits are rooted in a strong and...powerful will are not subject to contradiction. Repentance is but a denying of our will, and an opposition of our fantasies.

Princes give me sufficiently if they take nothing from me, and do me much good if they do me no hurt; it is all I require of them.

True it is that she who escapeth safe and unpolluted from out the school of freedom, giveth more confidence of herself than she who cometh sound out of the school of severity and restraint.

The greatest thing of the world is for a man to know how to be his own.

Few men have been admired of their familiars.

My reason is not framed to bend or stoop: my knees are.

Who feareth to suffer suffereth already, because he feareth.

It is the part of cowardliness, and not of virtue, to seek to squat itself in some hollow lurking hole, or to hide herself under some massive tomb, thereby to shun the strokes of fortune.

A man should ever…be ready booted to take his journey.

I speak truth, not my belly-full, but as much as I dare; and I dare the more the more I grow into years.

Virtue rejects facility to be her companion.…She requires a craggy, rough and thorny way.

Wisdom hath her excesses, and no less need of moderation than folly.

To philosophize is to learn to die.

Charles de Montesquieu

(1689-1755)
French philosopher, lawyer

If triangles made a god, they would give him three sides.

Lewis Mumford

(1895-1990)
U.S. social philosopher

Every new baby is a blind desperate vote for survival: people who find themselves unable to register an effective political protest against extermination do so by a biological act.

The city is a fact in nature, like a cave, a run of mackerel or an ant-heap. But it is also a conscious work of art, and it holds within its communal framework many simpler and more personal forms of art. Mind *takes form* in the city; and in turn, urban forms condition mind.

The chief function of the city is to convert power into form, energy into culture, dead matter into the living symbols of art, biological reproduction into social creativity.

We have created an industrial order geared to automatism, where feeble-mindedness, native or acquired, is necessary for docile productivity in the factory; and where a pervasive neurosis is the final gift of the meaningless life that issues forth at the other end.

The vast material displacements the machine has made in our physical environment are perhaps in the long run less important than its spiritual contributions to our culture.

The cycle of the machine is now coming to an end. Man has learned much in the hard discipline and the shrewd, unflinching grasp of practical possibilities that the machine has provided in the last three centuries: but we can no more continue to live in the world of the machine than we could live successfully on the barren surface of the moon.

Unable to create a meaningful life for itself, the personality takes its own revenge: from the lower depths comes a regressive form of spontaneity: raw animality forms a counterpoise to the meaningless stimuli and the vicarious life to which the ordinary man is conditioned. Getting spiritual nourishment from this chaos of events, sensations, and devious interpretations is the equivalent of trying to pick through a garbage pile for food.

The settlement of America had its origins in the unsettlement of Europe. America came into existence when the European was already so distant from the ancient ideas and ways of his birthplace that the whole span of the Atlantic did not widen the gulf.

Today, the degradation of the inner life is symbolized by the fact that the only place sacred from interruption is the private toilet.

Today, the notion of progress in a single line without goal or limit seems perhaps the most parochial notion of a very parochial century.

Sport in the sense of a mass-spectacle, with death to add to the underlying excitement, comes into existence when a population has been drilled and regimented and depressed to such an extent that it needs at least a vicarious participation in difficult feats of strength or skill or heroism in order to sustain its waning life-sense.

However far modern science and technics have fallen short of their inherent possibilities, they have taught mankind at least one lesson: Nothing is impossible.

By his very success in inventing labor-saving devices, modern man has manufactured an abyss of boredom that only the privileged classes in earlier civilizations have ever fathomed.

War is the supreme drama of a completely mechanized society.

Friedrich Nietzsche

(1844-1900)
German philosopher

In the consciousness of the truth he has perceived, man now sees everywhere only the awfulness or the absurdity of existence...and loathing seizes him.

The irrationality of a thing is no argument against its existence, rather a condition of it.

Nothing is beautiful, only man: on this piece of naïvety rests all aesthetics, it is the *first* truth of aesthetics. Let us immediately add its second: nothing is ugly but *degenerate* man—the domain of aesthetic judgment is therewith defined.

Where does one not find that bland degeneration which beer produces in the spirit!

For art to exist, for any sort of aesthetic activity or perception to exist, a certain physiological precondition is indispensable: *intoxication.*

The anarchist and the Christian have a common origin.

I fear animals regard man as a creature of their own kind which has in a highly dangerous fashion lost its healthy animal reason—as the mad animal, as the laughing animal, as the weeping animal, as the unhappy animal.

In the mountains the shortest route is from peak to peak, but for that you must have long legs. Aphorisms should be peaks: and those to whom they are spoken should be big and tall of stature.

The aphorism, the apophthegm, in which I am the first master among Germans, are the forms of "eternity"; my ambition is to say in ten sentences what everyone else says in a book—what everyone else *does not* say in a book.

The *architect* represents neither a Dionysian nor an Apollinian condition: here it is the mighty act of will, the will which moves mountains, the intoxication of the strong will, which demands artistic expression. The most powerful men have always inspired the architects; the architect has always been influenced by power.

One often contradicts an opinion when what is uncongenial is really the tone in which it was conveyed.

Art is not merely an imitation of the reality of nature, but in truth a metaphysical supplement to the reality of nature, placed alongside thereof for its conquest.

The ascetic makes a necessity of virtue.

In every ascetic morality man worships a part of himself as God and for that he needs to diabolize the other part.

In the beautiful, man sets himself up as the standard of perfection; in select cases he worships himself in it....Man believes that the world itself is filled with beauty—he *forgets* that it is he who has created it. He alone has bestowed beauty upon the world—alas! only a very human, an all too human, beauty.

Beggars...should be entirely abolished! Truly, it is annoying to give to them and annoying not to give to them.

The most dangerous follower is he whose defection would destroy the whole party: that is to say, the best follower.

Only the most acute and active animals are capable of boredom.—A theme for a great poet would be *God's boredom* on the seventh day of creation.

Against boredom the gods themselves fight in vain.

You say it is the good cause that hallows even war? I tell you: it is the good war that hallows every cause.

Two great European narcotics, alcohol and Christianity.

To exercise power costs effort and demands courage. That is why so many fail to assert rights to which they are perfectly entitled—because a right is a kind of *power* but they are too lazy or too cowardly to exercise it. The virtues which cloak these faults are called *patience* and *forbearance*.

A letter is an unannounced visit, the postman the agent of rude surprises. One ought to reserve an hour a week for receiving letters and afterwards take a bath.

The desire to create continually is vulgar and betrays jealousy, envy, ambition. If one is something one really does not need to make anything—and one nonetheless does very much. There exists above the "productive" man a yet higher species.

Almost everything we call "higher culture" is based on the spiritualization and intensification of *cruelty*—this is my proposition....That which constitutes the painful voluptuousness of tragedy is cruelty; that which produces a pleasing effect in so-called tragic pity, indeed fundamentally in everything sublime up to the most highest and most refined thrills of metaphysics, derives its sweetness solely from the ingredient of cruelty mixed in with it.

For believe me!—the secret of realizing the greatest fruitfulness and the greatest enjoyment of existence is: to *live dangerously*! Build your cities on the slopes of Vesuvius! Send your ships out into uncharted seas! Live in conflict with your equals and with yourselves! Be robbers and ravagers as long as you cannot be rulers and owners, you men of knowledge! The time will soon be past when you could be content to live concealed in the woods like timid deer!

The invalid is a parasite on society. In a certain state it is indecent to go on living. To vegetate on in cowardly dependence on physicians and medicaments after the meaning of life, the right to life, has been lost ought to entail the profound contempt of society.

We find nothing easier than being wise, patient, superior. We drip with the oil of forbearance and sympathy, we are absurdly just, we forgive everything. For that very reason we ought to discipline ourselves a little; for that very reason we ought to *cultivate* a little emotion, a little emotional vice, from time to time. It may be hard for us; and among ourselves we may perhaps laugh at the appearance we thus present. But what of that! We no longer have any other mode of self-overcoming available to us: this is *our* asceticism, *our* penance.

You may have enemies whom you hate, but not enemies whom you despise. You must be proud of your enemy: then the success of your enemy shall be your success too.

To die proudly when it is no longer possible to live proudly. Death of one's own free choice, death at the proper time, with a clear head and with joyfulness, consummated in the midst of children and witnesses: so that an actual leave-taking is possible while he who is leaving *is still there.*

Extreme positions are not succeeded by moderate ones, but by *contrary* extreme positions.

It says nothing against the ripeness of a spirit that it has a few worms.

Fanatics are picturesque, mankind would rather see gestures than listen to *reasons.*

If one considers how much reason every person has for anxiety and timid self-concealment, and how three-quarters of his energy and goodwill can be paralyzed and made unfruitful by it, one has to be very grateful to fashion, insofar as it sets that three-quarters free and communicates self-confidence and mutual cheerful agreeableness to those who know they are subject to its law.

I assess the power of a will by how much resistance, pain, torture it endures and knows how to turn to its advantage.

How is freedom measured, in individuals as in nations? By the resistance which has to be overcome, by the effort it costs to stay *aloft.* One would have to seek the highest type of free man where the greatest resistance is constantly being overcome: five steps from tyranny, near the threshold of the danger of servitude.

A woman may very well form a friendship with a man, but for this to endure, it must be assisted by a little physical antipathy.

Everything ponderous, viscous, and pompously clumsy, all long-winded and wearying species of style are developed in profuse variety among Germans.

How much dreary heaviness, lameness, dampness, sloppiness, how much *beer* there is in the German intellect!

There is in general good reason to suppose that in several respects the gods could all benefit from instruction by us human beings. We humans are—more humane.

What is good?—All that heightens the feeling of power, the will to power, power itself in man.

It was *modesty* which in Greece invented the word "philosopher" and left the splendid arrogance of calling oneself wise to the actors of the spirit—the modesty of such monsters of pride and self-glorification as Pythagoras, as Plato.

The man of knowledge must be able not only to love his enemies but also to hate his friends.

The "kingdom of Heaven" is a condition of the heart—not something that comes "upon the earth" or "after death."

To live alone one must be an animal or a god—says Aristotle. There is yet a third case: one must be both—a *philosopher*.

Only strong personalities can endure history, the weak ones are extinguished by it.

Man is no longer an artist, he has become a work of art.

The belly is the reason why man does not mistake himself for a god.

The idealist is incorrigible: if he is thrown out of his heaven he makes an ideal of his hell.

And as for sickness: would we not almost be tempted to ask whether we can in any way do without it? Only great pain is, as the teacher of *great suspicion*, the ultimate liberator of the spirit....It is only great pain, that slow protracted pain which takes its time and in which we are as it were burned with green wood, that compels us philosophers to descend into our ultimate depths and to put from us all trust, all that is good-hearted, palliated, gentle, average, wherein perhaps our humanity previously reposed. I doubt whether such pain "improves"—but I do know it *deepens* us.

We operate with nothing but things which do not exist, with lines, planes, bodies, atoms, divisible time, divisible space—how should explanation even be possible when we first make everything into an *image*, into our own image!

Instinct. When the house burns one forgets even lunch. Yes, but one eats it later in the ashes.

One receives as reward for much *ennui*, despondency, boredom—such as a solitude without friends, books, duties, passions must bring with it—those quarter-hours of profoundest contemplation within oneself and nature. He who completely entrenches himself against boredom also entrenches himself against himself: he will never get to drink the strongest refreshing draught from his own innermost fountain.

The word "Christianity" is already a misunderstanding—in reality there has been only one Christian, and he died on the Cross.

We have no organ at all for *knowledge*, for "truth": we "know" (or believe or imagine) precisely as much as may be *useful* in the interest of the human herd, the species: and even what is here called "usefulness" is in the end only a belief, something imagined and perhaps precisely that most fatal piece of stupidity by which we shall one day perish.

Our treasure lies in the beehive of our knowledge. We are perpetually on the way thither, being by nature winged insects and honey gatherers of the mind.

The significance of language for the evolution of culture lies in this, that mankind set up in language a separate world beside the other world, a place it took to be so firmly set that, standing upon it, it could lift the rest of the world off its hinges and make itself master of it. To the extent that man has for long ages believed in the concepts and names of things as in *aeternae veritates* he has appropriated to himself that pride by which he raised himself above the animal: he really thought that in language he possessed knowledge of the world.

No man lies so boldly as the man who is indignant.

Let us beware of saying that death is the opposite of life. The living being is only a species of the dead, and a very rare species.

The spiritualization of sensuality is called *love*: it is a great triumph over Christianity.

Madness is something rare in individuals—but in groups, parties, peoples, ages it is the rule.

The best friend is likely to acquire the best wife, because a good marriage is based on the talent for friendship.

Mathematics…would certainly have not come into existence if one had known from the beginning that there was in nature no exactly straight line, no actual circle, no absolute magnitude.

The true man wants two things: danger and play. For that reason he wants woman, as the most dangerous plaything.

For the woman, the man is a means: the end is always the child.

You will never get the crowd to cry Hosanna until you ride into town on an ass.

Let us beware of saying there are laws in nature. There are only necessities: there is no one to command, no one to obey, no one to transgress. When you realize there are no goals or objectives, then you realize, too, that there is no chance: for only in a world of objectives does the word "chance" have any meaning.

Not when truth is dirty, but when it is shallow, does the enlightened man dislike to wade into its waters.

Altered opinions do not alter a man's character (or do so very little); but they do illuminate individual aspects of the constellation of his personality which with a different constellation of opinions had hitherto remained dark and unrecognizable.

We would not let ourselves be burned to death for our opinions: we are not sure enough of them for that. But perhaps for the right to have our opinions and to change them.

Everyone who has ever built anywhere a "new heaven" first found the power thereto in his own hell.

Man…cannot learn to forget, but hangs on the past: however far or fast he runs, that chain runs with him.

Because men really respect only that which was founded of old and has developed slowly, he who wants to live on after his death must take care not only of his posterity but even more of his past.

The philosopher believes that the value of his philosophy lies in the whole, in the building: posterity discovers it in the bricks with which he built and which are then often used again for better building: in the fact, that is to say, that that building can be destroyed and *nonetheless* possess value as material.

Actual philosophers…are commanders and law-givers: they say "thus it shall be!", it is they who determine the Wherefore and Whither of mankind, and they possess for this task the preliminary work of all the philosophical labourers, of all those who have subdued the past—they reach for the future with creative hand, and everything that is or has been becomes for them a means, an instrument, a hammer. Their "knowing" is *creating*, their creating is a lawgiving, their will to truth is—*will to power*. Are their such philosophers today? Have there been such philosophers? *Must* there not be such philosophers?

We must be physicists in order...to be creative since so far codes of values and ideals have been constructed in ignorance of physics or even in contradiction to physics.

Not necessity, not desire—no, the love of power is the demon of men. Let them have everything—health, food, a place to live, entertainment—they are and remain unhappy and low-spirited: for the demon waits and waits and will be satisfied.

So long as you are praised think only that you are not yet on your own path but on that of another.

All in all, punishment hardens and renders people more insensible; it concentrates; it increases the feeling of estrangement; it strengthens the power of resistance.

Distrust everyone in whom the impulse to punish is powerful!

The worst readers are those who behave like plundering troops: they take away a few things they can use, dirty and confound the remainder, and revile the whole.

Early in the morning, at break of day, in all the freshness and dawn of one's strength, to read a *book*—I call that vicious!

After coming into contact with a religious man I always feel I must wash my hands.

Active, successful natures act, not according to the dictum "know thyself," but as if there hovered before them the commandment: *will* a self and thou shalt *become* a self.

He who cannot obey himself will be commanded. That is the nature of living creatures.

"Reason" is the cause of our falsification of the evidence of the senses. In so far as the senses show becoming, passing away, change, they do not lie.

How nicely the bitch Sensuality knows how to beg for a piece of the spirit, when a piece of flesh is denied her.

The most spiritual human beings, assuming they are the most courageous, also experience by far the most painful tragedies: but it is precisely for this reason that they honor life, because it brings against them its most formidable weapons.

"To give style" to one's character—a great and rare art! He exercises it who surveys all that his nature presents in strength and weakness and then moulds it to an artistic plan until everything appears as art and reason, and even the weaknesses delight the eye.

It is always consoling to think of suicide: in that way one gets through many a bad night.

Reckoned physiologically, everything ugly weakens and afflicts man. It recalls decay, danger, impotence; he actually suffers a loss of energy in its presence. The effect of the ugly can be measured with a dynamometer. Whenever man feels in any way depressed, he senses the proximity of something "ugly." His feeling of power, his will to power, his courage, his pride—they decline with the ugly, they increase with the beautiful.

We do not place especial value on the possession of a virtue until we notice its total absence in our opponent.

All good things were at one time bad things; every original sin has developed into an original virtue.

War has always been the grand sagacity of every spirit which has grown too inward and too profound; its curative power lies even in the wounds one receives.

Does wisdom perhaps appear on the earth as a raven which is inspired by the smell of carrion?

Women are considered deep—why? Because one can never discover any bottom to them. Women are not even shallow.

If a woman possesses manly virtues one should run away from her; and if she does not possess them she runs away from herself.

"Stupid as a man" say the women: "cowardly as a woman" say the men. Stupidity is in woman the *unwomanly*.

José Ortega y Gasset

(1883-1955)
Spanish essayist, philosopher

Were art to redeem man, it could do so only by saving him from the seriousness of life and restoring him to an unexpected boyishness.

Being an artist means ceasing to take seriously that very serious person we are when we are not an artist.

Barbarism is the absence of standards to which appeal can be made.

Biography is: a system in which the contradictions of a human life are unified.

There is but one way left to save a classic: to give up revering him and use him for our own salvation.

In order to master the unruly torrent of life the learned man meditates, the poet quivers, and the political hero erects the fortress of his will.

The characteristic of the hour is that the commonplace mind, knowing itself to be commonplace, has the assurance to proclaim the rights of the commonplace and to impose them wherever it will.

Towns are full of people, houses full of tenants, hotels full of guests, trains full of travelers, cafés full of customers, parks full of promenaders, consulting-rooms of famous doctors full of patients, theatres full of spectators, and beaches full of bathers. What previously was, in general, no problem, now begins to be an everyday one, namely, to find room.

The cynic, a parasite of civilisation, lives by denying it, for the very reason that he is convinced that it will not fail.

By speaking, by thinking, we undertake to clarify things, and that forces us to exacerbate them, dislocate them, schematize them. Every concept is in itself an exaggeration.

The essence of man is, discontent, divine discontent; a sort of love without a beloved, the ache we feel in a member we no longer have.

Effort is only effort when it begins to hurt.

I am I plus my surroundings and if I do not preserve the latter, I do not preserve myself.

Better beware of notions like genius and inspiration; they are a sort of magic wand and should be used sparingly by anybody who wants to see things clearly.

The good is, like nature, an immense landscape in which man advances through centuries of exploration.

To rule is not so much a question of the heavy hand as the firm seat.

Hatred is a feeling which leads to the extinction of values.

We have need of history in its entirety, not to fall back into it, but to see if we can escape from it.

Liberalism—it is well to recall this today—is the supreme form of generosity; it is the right which the majority concedes to minorities and hence it is *the noblest cry* that has ever resounded in this planet. It announces the determination to share existence with the enemy; more than that, with an enemy which is weak.

Life is a petty thing unless it is moved by the indominatable urge to extend its boundaries. Only in proportion as we are desirous of living more do we really live.

Life is an operation which is done in a forward direction. One lives *toward* the future, because to live consists inexorably in *doing*, in each individual life *making* itself.

There may be as much nobility in being last as in being first, because the two positions are equally necessary in the world, the one to complement the other.

The mass believes that it has the right to impose and to give force of law to notions born in the café.

Poetry has become the higher algebra of metaphors.

Poetry is adolescence fermented, and thus preserved.

The poet begins where the man ends. The man's lot is to live his human life, the poet's to invent what is nonexistent.

I do not deny that there may be other well-founded causes for the hatred which various classes feel toward politicians, but the *main one seems to me that politicians are symbols of the fact that every class must take every other class into account.*

A revolution does not last more than fifteen years, the period which coincides with the flourishing of a generation.

Stupefaction, when it persists, becomes stupidity.

He who wishes to teach us a truth should not tell it to us, but simply suggest it with a brief gesture, a gesture which starts an ideal trajectory in the air along which we glide until we find ourselves at the feet of the new truth.

For the person for whom small things do not exist, the great is not great.

We do not live to think, but, on the contrary, we think in order that we may succeed in surviving.

An "unemployed" existence is a worse negation of life than death itself.

Youth does not require reasons for living, it only needs pretexts.

Blaise Pascal

(1623-62)
French scientist, philosopher

Animals do not admire each other. A horse does not admire its companion.

The last thing one discovers in composing a work is what to put first.

Man finds nothing so intolerable as to be in a state of complete rest, without passions, without occupation, without diversion, without effort. Then he feels his nullity, loneliness, inadequacy, dependence, helplessness, emptiness.

It is superstitious to put one's hopes in formalities, but arrogant to refuse to submit to them.

Men never do evil so fully and cheerfully as when we do it out of conscience.

Man is only a reed, the weakest in nature; but he is a thinking reed. There is no need for the whole universe to take up arms to crush him: a vapor, a drop of water is enough to kill him. But even if the universe were to crush him, man would still be nobler than his slayer, because he knows that he is dying and the advantage the universe has over him. The universe knows nothing of this.

Faith certainly tells us what the senses do not, but not the contrary of what they see; it is above, not against them.

It is the heart which perceives God and not the reason. That is what faith is: God perceived by the heart, not by the reason.

I maintain that, if everyone knew what others said about him, there would not be four friends in the world.

We like security: we like the pope to be infallible in matters of faith, and grave doctors to be so in moral questions so that we can feel reassured.

The more intelligent one is, the more men of originality one finds. Ordinary people find no difference between men.

The heart has its reasons of which reason knows nothing: we know this in countless ways.

Men despise religion. They hate it and are afraid it may be true.

Vanity of science. Knowledge of physical science will not console me for ignorance of morality in time of affliction, but knowledge of morality will always console me for ignorance of physical science.

We run heedlessly into the abyss after putting something in front of us to stop us seeing it.

There are only two kinds of men: the righteous who think they are sinners and the sinners who think they are righteous.

Man is obviously made for thinking. Therein lies all his dignity and his merit; and his whole duty is to think as he ought.

All our dignity consists in thought. It is on thought that we must depend for our reovery, not on space and time which we could never fill. Let us then strive to think well; that is the basic principle of morality.

When we see a natural style we are quite amazed and delighted, because we expected to see an author and find a man.

Jan Patocka

(1907-77)
Czech philosopher, activist

The real test of a man is not how well he plays the role he has invented for himself, but how well he plays the role that destiny assigned to him.

Philo of Alexander

(c.20 B.C.-A.D. c.50)
Philosopher and diplomat

Equality is the mother of Justice, queen of all virtues.

Plato

(427?-347? B.C.)
Greek philosopher

It is clear to everyone that astronomy at all events compels the soul to look upwards, and draws it from the things of this world to the other.

These, then, will be some of the features of democracy…it will be, in all likelihood, an agreeable, lawless, particolored commonwealth, dealing with all alike on a footing of equality, whether they be really equal or not.

Is it not also true that no physician, in so far as he is a physician, considers or enjoins what is for the physician's interest, but that all seek the good of their patients? For we have agreed that a physician strictly so called, is a ruler of bodies, and not a maker of money, have we not?

Let us describe the education of our men.…What then is the education to be? Perhaps we could hardly find a better than that which the experience of the past has already discovered, which consists, I believe, in gymnastic, for the body, and music for the mind.

In the world of knowledge, the essential Form of Good is the limit of our inquiries, and can barely be perceived; but, when perceived, we cannot help concluding that it is in every case the source of all that is bright and beautiful—in the visible world giving birth to light and its master, and in the intellectual world dispensing, immediately and with full authority, truth and reason—and that whosoever would act wisely, either in private or in public, must set this Form of Good before his eyes.

The punishment which the wise suffer who refuse to take part in the government, is to live under the government of worse men.

To the rulers of the state then, if to any, it belongs of right to use falsehood, to deceive either enemies or their own citizens, for the good of the state: and no one else may meddle with this privilege.

I have hardly ever known a mathematician who was capable of reasoning.

He who is of a calm and happy nature will hardly feel the pressure of age, but to him who is of an opposite disposition youth and age are equally a burden.

Poets utter great and wise things which they do not themselves understand.

All things will be produced in superior quantity and quality, and with greater ease, when each man works at a single occupation, in accordance with his natural gifts, and at the right moment, without meddling with anything else.

Whenever a person strives, by the help of dialectic, to start in pursuit of every reality by a simple process of reason, independent of all sensuous information—never flinching, until by an act of the pure intelligence he has grasped the real nature of good—he arrives at the very end of the intellectual world.

For just as poets love their own works, and fathers their own children, in the same way those who have created a fortune value their money, not merely for its uses, like other persons, but because it is their own production. This makes them moreover disagreeable companions, because they will praise nothing but riches.

We ought to esteem it of the greatest importance that the fictions which children first hear should be adapted in the most perfect manner to the promotion of virtue.

You must train the children to their studies in a playful manner, and without any air of constraint, with the further object of discerning more readily the natural bent of their respective characters.

Thomas Reid

(1710-69)
Scottish philosopher

There is no greater impediment to the advancement of knowledge than the ambiguity of words.

Jean-Jacques Rousseau

(1712-78)
Swiss-born French philosopher, political theorist

We are born, so to speak, twice over; born into existence, and born into life; born a human being, and born a man.

Let the trumpet of the day of judgment sound when it will, I shall appear with this book in my hand before the Sovereign Judge, and cry with a loud voice, This is my work, there were my thoughts, and thus was I. I have freely told both the good and the bad, have hid nothing wicked, added nothing good.

We are born weak, we need strength; helpless, we need aid; foolish, we need reason. All that we lack at birth, all that we need when we come to man's estate, is the gift of education.

The English people believes itself to be free; it is gravely mistaken; it is free only during election of members of parliament; as soon as the members are elected, the people is enslaved; it is nothing. In the brief moment of its freedom, the English people makes such a use of that freedom that it deserves to lose it.

Man is born free, and everywhere he is in chains.

Reading, solitude, idleness, a soft and sedentary life, intercourse with women and young people, these are perilous paths for a young man, and these lead him constantly into danger.

Although modesty is natural to man, it is not natural to children. Modesty only begins with the knowledge of evil.

There are two things to be considered with regard to any scheme. In the first place, "Is it good in itself?" In the second, "Can it be easily put into practice?"

The first step towards vice is to shroud innocent actions in mystery, and whoever likes to conceal something sooner or later has reason to conceal it.

It is too difficult to think nobly when one thinks only of earning a living.

Bertrand Russell

(1872-1970)
British philosopher, mathematician

Drunkenness…is temporary suicide.

Indignation is a submission of our thoughts, but not of our desires.

When the intensity of emotional conviction subsides, a man who is in the habit of reasoning will search for logical grounds in favour of the belief which he finds in himself.

Boredom is…a vital problem for the moralist, since at least half the sins of mankind are caused by the fear of it.

Advocates of capitalism are very apt to appeal to the sacred principles of liberty, which are embodied in one maxim: *The fortunate must not be restrained in the exercise of tyranny over the unfortunate.*

If all our happiness is bound up entirely in our personal circumstances it is difficult not to demand of life more than it has to give.

Conventional people are roused to fury by departures from convention, largely because they regard such departures as a criticism of themselves.

Man is a credulous animal, and must believe *something*; in the absence of good grounds for belief, he will be satisfied with bad ones.

A sense of duty is useful in work but offensive in personal relations. People wish to be liked, not to be endured with patient resignation.

To acquire immunity to eloquence is of the utmost importance to the citizens of a democracy.

In America everybody is of opinion that he has no social superiors, since all men are equal, but he does not admit that he has no social inferiors.

Ethics is in origin the art of recommending to others the sacrifices required for cooperation with oneself.

Organic life, we are told, has developed gradually from the protozoan to the philosopher, and this development, we are assured, is indubitably an advance. Unfortunately it is the philosopher, not the protozoan, who gives us this assurance.

An extra-terrestrial philosopher, who had watched a single youth up to the age of twenty-one and had never come across any other human being, might conclude that it is the nature of human beings to grow continually taller and wiser in an indefinite progress towards perfection; and this generalisation would be just as well founded as the generalisation which evolutionists base upon the previous history of this planet.

In the revolt against idealism, the ambiguities of the word "experience" have been perceived, with the result that realists have more and more avoided the word.

Those who forget good and evil and seek only to know the facts are more likely to achieve good than those who view the world through the distorting medium of their own desires.

With the introduction of agriculture mankind entered upon a long period of meanness, misery, and madness, from which they are only now being freed by the beneficent operation of the machine.

A hallucination is a fact, not an error; what is erroneous is a judgment based upon it.

If any philosopher had been asked for a definition of infinity, he might have produced some unintelligible rigmarole, but he would certainly not have been able to give a definition that had any meaning at all.

To be able to fill leisure intelligently is the last product of civilization.

The essence of the Liberal outlook lies not in *what* opinions are held, but in *how* they are held: instead of being held dogmatically, they are held tentatively, and with a consciousness that new evidence may at any moment lead to their abandonment.

The life of man is a long march through the night, surrounded by invisible foes, tortured by weariness and pain, towards a goal that few can hope to reach, and where none may tarry long.

Many people when they fall in love look for a little haven of refuge from the world, where they can be sure of being admired when they are not admirable, and praised when they are not praiseworthy.

Machines are worshipped because they are beautiful and valued because they confer power; they are hated because they are hideous and loathed because they impose slavery.

Marriage is for women the commonest mode of livelihood, and the total amount of undesired sex endured by women is probably greater in marriage than in prostitution.

Freedom comes only to those who no longer ask of life that it shall yield them any of those personal goods that are subject to the mutations of time.

Mathematics may be defined as the subject in which we never know what we are talking about, nor whether what we are saying is true.

I like mathematics because it is not human and has nothing particular to do with this planet or with the whole accidental universe—because, like Spinoza's God, it won't love us in return.

There is no need to worry about mere size. We do not necessarily respect a fat man more than a thin man. Sir Isaac Newton was very much smaller than a hippopotamus, but we do not on that account value him less.

This idea of weapons of mass extermination is utterly horrible and is something which no one with one spark of humanity can tolerate. I will not pretend to obey a government which is organizing a mass massacre of mankind.

Nine-tenths of the appeal of pornography is due to the indecent feelings concerning sex which moralists inculcate in the young; the other tenth is physiological, and will occur in one way or another whatever the state of the law may be.

The fundamental concept in social science is Power, in the same sense in which Energy is the fundamental concept in physics.

A process which led from the amoeba to man appeared to the philosophers to be obviously a progress—though whether the amoeba would agree with this opinion is not known.

Why is propaganda so much more successful when it stirs up hatred than when it tries to stir up friendly feeling?

Reason is a harmonising, controlling force rather than a creative one.

Religions, which condemn the pleasures of sense, drive men to seek the pleasures of power. Throughout history power has been the vice of the ascetic.

In science men have discovered an activity of the very highest value in which they are no longer, as in art, dependent for progress upon the appearance of continually greater genius, for in science the successors stand upon the shoulders of their predecessors; where one man of supreme genius has invented a method, a thousand lesser men can apply it.

Can a society in which thought and technique are scientific persist for a long period, as, for example, ancient Egypt persisted, or does it necessarily contain within itself forces which must bring either decay or explosion?

Aristotle could have avoided the mistake of thinking that women have fewer teeth than men, by the simple device of asking Mrs. Aristotle to keep her mouth open while he counted.

The slave is doomed to worship time and fate and death, because they are greater than anything he finds in himself, and because all his thoughts are of things which they devour.

A truer image of the world, I think, is obtained by picturing things as entering into the stream of time from an eternal world outside, than from a view which regards time as the devouring tyrant of all that is.

The theoretical understanding of the world, which is the aim of philosophy, is not a matter of great practical importance to animals, or to savages, or even to most civilised men.

Men who are unhappy, like men who sleep badly, are always proud of the fact.

I had supposed until that time that it was quite common for parents to love their children, but the war persuaded me that it is a rare exception. I had supposed that most people liked money better than almost anything else, but I discovered that they liked destruction even better. I had supposed that intellectuals frequently loved truth, but I found here again that not ten per cent of them prefer truth to popularity.

For my part I distrust *all* generalizations about women, favourable and unfavourable, masculine and feminine, ancient and modern; all alike, I should say, result from paucity of experience.

Admiration of the proletariat, like that of dams, power stations, and aeroplanes, is part of the ideology of the machine age.

George Santayana

(1863-1952)
U.S. philosopher, poet

To knock a thing down, especially if it is cocked at an arrogant angle, is a deep delight to the blood.

The effort of art is to keep what is interesting in existence, to recreate it in the eternal.

The body is an instrument, the mind its function, the witness and reward of its operation.

Chaos is a name for any order that produces confusion in our minds.

Character is the basis of happiness and happiness the sanction of character.

The primary use of conversation is to satisfy the impulse to talk.

Perhaps the only true dignity of man is his capacity to despise himself.

The young man who has not wept is a savage, and the old man who will not laugh is a fool.

Fanaticism consists in redoubling your effort when you have forgotten your aim.

Fashion is something barbarous, for it produces innovation without reason and imitation without benefit.

That fear first created the gods is perhaps as true as anything so brief could be on so great a subject.

A man's feet must be planted in his country, but his eyes should survey the world.

Fun is a good thing but only when it spoils nothing better.

Happiness is the only sanction of life; where happiness fails, existence remains a mad and lamentable experiment.

Those who cannot remember the past are condemned to repeat it.

It takes patience to appreciate domestic bliss; volatile spirits prefer unhappiness.

The more rational an institution is the less it suffers by making concessions to others.

Nothing can be meaner than the anxiety to live on, to live on anyhow and in any shape; a spirit with any honor is not willing to live except in its own way, and a spirit with any wisdom is not over-eager to live at all.

There is no cure for birth and death save to enjoy the interval.

When men and women agree, it is only in their conclusions; their reasons are always different.

Parents lend children their experience and a vicarious memory; children endow their parents with a vicarious immortality.

Many possessions, if they do not make a man better, are at least expected to make his children happier; and this pathetic hope is behind many exertions.

Progress, far from consisting in change, depends on retentiveness. When change is absolute there remains no being to improve and no direction is set for possible improvement: and when experience is not retained, as among savages, infancy is perpetual.

The irrational in the human has something about it altogether repulsive and terrible, as we see in the maniac, the miser, the drunkard or the ape.

The mind of the Renaissance was not a pilgrim mind, but a sedentary city mind, like that of the ancients.

It is easier to make a saint out of a libertine than out of a prig.

Sanity is a madness put to good uses; waking life is a dream controlled.

To be interested in the changing seasons is, in this middling zone, a happier state of mind than to be hopelessly in love with spring.

There is a kind of courtesy in skepticism. It would be an offense against polite conventions to press our doubts too far.

Society is like the air, necessary to breathe but insufficient to live on.

A soul is but the last bubble of a long fermentation in the world.

The spirit's foe in man has not been simplicity, but sophistication.

Oaths are the fossils of piety.

The theatre, for all its artifices, depicts life in a sense more truly than history, because the medium has a kindred movement to that of real life, though an artificial setting and form.

A conception not reducible to the small change of daily experience is like a currency not exchangeable for articles of consumption; it is not a symbol, but a fraud.

America is a young country with an old mentality.

Oxford, the paradise of dead philosophies.

To delight in war is a merit in the soldier, a dangerous quality in the captain, and a positive crime in the statesman.

The world is a perpetual caricature of itself; at every moment it is the mockery and the contradiction of what it is pretending to be.

Wealth, religion, military victory have more rhetorical than efficacious worth.

Jean-Paul Sartre

(1905-80)
French philosopher, author

What then did you expect when you unbound the gag that muted those black mouths? That they would chant your praises? Did you think that when those heads that our fathers had forcibly bowed down to the ground were raised again, you would find adoration in their eyes?

One is still what one is going to cease to be and already what one is going to become. One lives one's death, one dies one's life.

Fascism is not defined by the number of its victims, but by the way it kills them.

To eat is to appropriate by destruction.

Generosity is nothing else than a craze to possess. All which I abandon, all which I give, I enjoy in a higher manner through the fact that I give it away....To give is to enjoy possessively the object which one gives.

Man is a useless passion.

If literature isn't *everything*, it's not worth a single hour of someone's trouble.

Hell is other people.

Every age has its own poetry; in every age the circumstances of history choose a nation, a race, a class to take up the torch by creating situations that can be expressed or transcended only through poetry.

I am responsible for everything...except for my very responsibility, for I am not the foundation of my being. Therefore everything takes place as if I were compelled to be responsible. I am *abandoned* in the world...in the sense that I find myself suddenly alone and without help, engaged in a world for which I bear the whole responsibility without being able, whatever I do, to tear myself away from this responsibility for an instant.

Friedrich Schlegel

(1772-1829)
German philosopher, critic, writer

An aphorism ought to be entirely isolated from the surrounding world like a little work of art and complete in itself like a hedgehog.

All men are somewhat ridiculous and grotesque; just because they are men; and in this respect, artists might well be regarded as man multiplied by two. So it is, was, and shall be.

A critic is a reader who ruminates. Thus, he should have more than one stomach.

Every uneducated person is a caricature of himself.

Genius is, to be sure, not a matter of arbitariness, but rather of freedom, just as wit, love, and faith, which once shall become arts and disciplines. We should demand genius from everybody, without, however, expecting it.

The subject of history is the gradual realization of all that is practically necessary.

It is peculiar to mankind to transcend mankind.

Prudishness is pretense of innocence without innocence. Women have to remain prudish as long as men are sentimental, dense, and evil enough to demand of them eternal innocence and lack of education. For innocence is the only thing which can ennoble lack of education.

What is called good society is usually nothing but a mosaic of polished caricatures.

Good drama must be drastic.

Virtue is reason which has become energy.

Arthur Schopenhauer

(1788-1860)
German philosopher

In the sphere of thought, absurdity and perversity remain the masters of the world, and their dominion is suspended only for brief periods.

There is no doubt that life is given us, not to be enjoyed, but to be overcome—to be got over.

The brain may be regarded as a kind of parasite of the organism, a pensioner, as it were, who dwells with the body.

Buying books would be a good thing if one could also buy the time to read them in: but as a rule the purchase of books is mistaken for the appropriation of their contents.

Because people have no thoughts to deal in, they deal cards, and try and win one another's money. Idiots!

All the cruelty and torment of which the world is full is in fact merely the necessary result of the totality of the forms under which the will to live is objectified.

Each day is a little life: every waking and rising a little birth, every fresh morning a little youth, every going to rest and sleep a little death.

We can come to look upon the deaths of our enemies with as much regret as we feel for those of our friends, namely, when we miss their existence as witnesses to our success.

A man's face as a rule says more, and more interesting things, than his mouth, for it is a compendium of everything his mouth will ever say, in that it is the monogram of all this man's thoughts and aspirations.

Every parting gives a foretaste of death, every reunion a hint of the resurrection.

Great minds are related to the brief span of time during which they live as great buildings are to a little square in which they stand: you cannot see them in all their magnitude because you are standing too close to them.

Hatred is an affair of the heart; *contempt* that of the head.

Honor has not to be won; it must only not be lost.

Nature shows that with the growth of intelligence comes increased capacity for pain, and it is only with the highest degree of intelligence that suffering reaches its supreme point.

As the biggest library if it is in disorder is not as useful as a small but well-arranged one, so you may accumulate a vast amount of knowledge but it will be of far less value to you than a much smaller amount if you have not thought it over for yourself.

The word of man is the most durable of all material.

In our monogamous part of the world, to marry means to halve one's rights and double one's duties.

Money is human happiness in the abstract: he, then, who is no longer capable of enjoying human happiness in the concrete devotes his heart entirely to money.

National character is only another name for the particular form which the littleness, perversity and baseness of mankind take in every country. Every nation mocks at other nations, and all are right.

Newspapers are the second hand of history. This hand, however, is usually not only of inferior metal to the other hands, it also seldom works properly.

Obstinacy is the result of the will forcing itself into the place of the intellect.

Patriotism, when it wants to make itself felt in the domain of learning, is a dirty fellow who should be thrown out of doors.

How very paltry and limited the normal human intellect is, and how little lucidity there is in the human consciousness, may be judged from the fact that, despite the ephemeral brevity of human life, the uncertainty of our existence and the countless enigmas which press upon us from all sides, everyone does not continually and ceaselessly philosophize, but that only the rarest of exceptions do.

That the outer man is a picture of the inner, and the face an expression and revelation of the whole character, is a presumption likely enough in itself, and therefore a safe one to go on; borne out as it is by the fact that people are always anxious to see anyone who has made himself famous....Photography...offers the most complete satisfaction of our curiosity.

The discovery of truth is prevented more effectively, not by the false appearance things present and which mislead into error, not directly by weakness of the reasoning powers, but by preconceived opinion, by prejudice.

Reading is equivalent to thinking with someone else's head instead of with one's own.

Just as the largest library, badly arranged, is not so useful as a very moderate one that is well arranged, so the greatest amount of knowledge, if not elaborated by our own thoughts, is worth much less than a far smaller volume that has been abundantly and repeatedly thought over.

Rascals are always sociable—more's the pity! and the chief sign that a man has any nobility in his character is the little pleasure he takes in others' company.

Suicide may also be regarded as an experiment—a question which man puts to Nature, trying to force her to answer. The question is this: What change will death produce in a man's existence and in his insight into the nature of things? It is a clumsy experiment to make; for it involves the destruction of the very consciousness which puts the question and awaits the answer.

The fundament upon which all our knowledge and learning rests is the inexplicable.

Only a male intellect clouded by the sexual drive could call the stunted, narrow-shouldered, broad-hipped and short-legged sex the fair sex.

Seneca

(c.3 B.C.-A.D. 65)
Roman writer, philosopher, statesman

The bravest sight in the world is to see a great man struggling against adversity.

Drunkenness is nothing but voluntary madness.

The final hour when we cease to exist does not itself bring death; it merely of itself completes the death-process. We reach death at that moment, but we have been a long time on the way.

The evil which assails us is not in the localities we inhabit but in ourselves. We lack strength to endure the least task, being incapable of suffering pain, powerless to enjoy pleasure, impatient with everything. How many invoke death when, after having tried every sort of change, they find themselves reverting to the same sensations, unable to discover any new experience.

Just as I shall select my ship when I am about to go on a voyage, or my house when I propose to take a residence, so I shall choose my death when I am about to depart from life.

It is the superfluous things for which men sweat.

Nothing becomes so offensive so quickly as grief. When fresh it finds someone to console it, but when it becomes chronic, it is ridiculed, and rightly.

The display of grief makes more demands than grief itself. How few men are sad in their own company.

Most men ebb and flow in wretchedness between the fear of death and the hardship of life; they are unwilling to live, and yet they do not know how to die.

In my own time there have been inventions of this sort, transparent windows...tubes for diffusing warmth equally through all parts of a building...short-hand, which has been carried to such a perfection that a writer can keep pace with the most rapid speaker. But the inventing of such things is drudgery for the lowest slaves; philosophy lies deeper. It is not her office to teach men how to use their hands. The object of her lessons is to form the soul.

That moderation which nature prescribes, which limits our desires by resources restricted to our needs, has abandoned the field; it has now come to this—that to want only what is enough is a sign both of boorishness and of utter destitution.

Pain, scorned by yonder gout-ridden wretch, endured by yonder dyspeptic in the midst of his dainties, borne bravely by the girl in travail. Slight thou art, if I can bear thee, short thou art if I cannot bear thee!

Whatever is well said by another, is mine.

We often want one thing and pray for another, not telling the truth even to the gods.

A large part of mankind is angry not with the sins, but with the sinners.

It makes a great deal of difference whether one wills not to sin or has not the knowledge to sin.

What difference does it make how much you have? What you do not have amounts to much more.

Socrates

(469-399 B.C.)
Greek philosopher

To fear death, my friends, is only to think ourselves wise, without being wise: for it is to think that we know what we do not know. For anything that men can tell, death may be the greatest good that can happen to them: but they fear it as if they knew quite well that it was the greatest of evils. And what is this but that shameful ignorance of thinking that we know what we do not know?

Whenever, therefore, people are deceived and form opinions wide of the truth, it is clear that the error has slid into their minds through the medium of certain resemblances to that truth.

Wars and revolutions and battles are due simply and solely to the body and its desires. All wars are undertaken for the acquisition of wealth; and the reason why we have to acquire wealth is the body, because we are slaves in its service.

We are in fact convinced that if we are ever to have pure knowledge of anything, we must get rid of the body and contemplate things by themselves with the soul by itself. It seems, to judge from the argument, that the wisdom which we desire and upon which we profess to have set our hearts will be attainable only when we are dead and not in our lifetime.

When desire, having rejected reason and overpowered judgment which leads to right, is set in the direction of the pleasure which beauty can inspire, and when again under the influence of its kindred desires it is moved with violent motion towards the beauty of corporeal forms, it acquires a surname from this very violent motion, and is called love.

A system of morality which is based on relative emotional values is a mere illusion, a thoroughly vulgar conception which has nothing sound in it and nothing true.

Ordinary people seem not to realize that those who really apply themselves in the right way to philosophy are directly and of their own accord preparing themselves for dying and death.

So I soon made up my mind about the poets too: I decided that it was not wisdom that enabled them to write their poetry, but a kind of instinct or inspiration, such as you find in seers and prophets who deliver all their sublime messages without knowing in the least what they mean.

I only wish that ordinary people had an unlimited capacity for doing harm; then they might have an unlimited power for doing good.

In every one of us there are two ruling and directing principles, whose guidance we follow wherever they may lead; the one being an innate desire of pleasure; the other, an acquired judgment which aspires after excellence.

Where there is reverence there is fear, but there is not reverence everywhere that there is fear, because fear presumably has a wider extension than reverence.

Every pleasure or pain has a sort of rivet with which it fastens the soul to the body and pins it down and makes it corporeal, accepting as true whatever the body certifies.

I was afraid that by observing objects with my eyes and trying to comprehend them with each of my other senses I might blind my soul altogether.

The unexamined life was not worth living.

Herbert Spencer

(1820-1903)
English philosopher

A living thing is distinguished from a dead thing by the multiplicity of the changes at any moment taking place in it.

The ultimate result of shielding men from the effects of folly, is to fill the world with fools.

The preservation of health is a *duty*. Few seem conscious that there is such a thing as physical morality.

Hero-worship is strongest where there is least regard for human freedom.

The Republican form of government is the highest form of government; but because of this it requires the highest type of human nature—a type nowhere at present existing.

Divine right of kings means the divine right of anyone who can get uppermost.

Science is organized knowledge.

People are beginning to see that the first requisite to success in life is to be a good animal.

Benedict Spinoza

(1632-77)
Dutch philosopher

I believe democracy to be of all forms of government the most natural, and the most constant with individual liberty. In it no one transfers his natural rights so absolutely that he has no further voice in affairs.

Statesman are suspected of plotting against mankind, rather than consulting their interests, and are esteemed more crafty than learned.

Everyone has as much right as he has might.

The ultimate aim of government is to free every man from fear, that he may live in all possible security. In fact, the true aim of government is liberty.

The most tyrannical governments are those which make crimes of opinions, for everyone has an inalienable right to his thoughts.

Freedom is absolutely necessary for progress in science and the liberal arts.

Nature abhors a vacuum.

God and all the attributes of God are equal.

He who would distinguish the true from the false must have an adequate idea of what is true and false.

Will and intellect are one and the same.

Surely human affairs would be far happier if the power in men to be silent were the same as that to speak. But experience more than sufficiently teaches that men govern nothing with more difficulty than their tongues.

Those who are believed to be most abject and humble are usually most ambitious and envious.

Man is a social animal.

All excellent things are as difficult as they are rare.

Henry David Thoreau

(1817-62)
U.S. philosopher, author, naturalist, essayist

He is the best sailor who can steer within fewest points of the wind, and exact a motive power out of the greatest obstacles.

I did not wish to take a cabin passage, but rather to go before the mast and on the deck of the world, for there I could best see the moonlight amid the mountains. I do not wish to go below now.

We seem but to linger in manhood to tell the dreams of our childhood, and they vanish out of memory ere we learn the language.

I have lived some thirty-odd years on this planet, and I have yet to hear the first syllable of valuable or even earnest advice from my seniors.

Oh, one world at a time!

As for doing good, that is one of the professions which are full. Moreover, I have tried it fairly, and…am satisfied that it does not agree with my constitution.

What is a country without rabbits and partridges? They are among the most simple and indigenous animal products; ancient and venerable families known to antiquity as to modern times; of the very hue and substance of Nature, nearest allied to leaves and to the ground.

True, there are architects so called in this country, and I have heard of one at least possessed with the idea of making architectural ornaments have a core of truth, a necessity, and hence a beauty, as if it were a revelation to him. All very well perhaps from his point of view, but only a little better than the common dilettantism.

After all the field of battle possesses many advantages over the drawing-room. There at least is no room for pretension or excessive ceremony, no shaking of hands or rubbing of noses, which make one doubt your sincerity, but hearty as well as hard hand-play. It at least exhibits one of the faces of humanity, the former only a mask.

On the death of a friend, we should consider that the fates through confidence have devolved on us the task of a double living, that we have henceforth to fulfill the promise of our friend's life also, in our own, to the world.

We feel at first as if some opportunities of kindness and sympathy were lost, but learn afterward that any *pure grief* is ample recompense for all. That is, if we are faithful;—for a spent grief is but sympathy with the soul that disposes events, and is as natural as the resin of Arabian trees.—Only nature has a right to grieve perpetually, for she only is innocent. Soon the ice will melt, and the blackbirds sing along the river which he frequented, as pleasantly as ever. The same everlasting serenity will appear in this face of God, and we will not be sorrowful, if he is not.

For eighteen hundred years, though perchance I have no right to say it, the New Testament has been written; yet where is the legislator who has wisdom and practical talent enough to avail himself of the light which it sheds on the science of legislation?

Every man is the builder of a temple, called his body, to the god he worships, after a style purely his own, nor can he get off by hammering marble instead. We are all sculptors and painters, and our material is our own flesh and blood and bones.

Books, not which afford us a cowering enjoyment, but in which each thought is of unusual daring; such as an idle man cannot read, and a timid one would not be entertained by, which even make us dangerous to existing institution—such call I good books.

How many a man has dated a new era in his life from the reading of a book! The book exists for us, perchance, that will explain our miracles and reveal new ones. The at present unutterable things we may find somewhere uttered.

For what are the classics but the noblest thoughts of man? They are the only oracles which are not decayed, and there are such answers to the most modern inquiry in them as Delphi and Dodona never gave. We might as well omit to study Nature because she is old.

It is difficult to begin without borrowing, but perhaps it is the most generous course thus to permit your fellow-men to have an interest in your enterprise.

If I seem to boast more than is becoming, my excuse is that I brag for humanity rather than for myself.

Pity the man who has a character to support—it is worse than a large family—he is silent poor indeed.

The generative energy, which, when we are loose, dissipates and makes us unclean, when we are continent invigorates and inspires us. Chastity is the flowering of man; and what are called Genius, Heroism, Holiness, and the like, are but various fruits which succeed it.

Why level downward to our dullest perception always, and praise that as common sense? The commonest sense is the sense of men asleep, which they express by snoring.

I have a great deal of company in my house; especially in the morning, when nobody calls.

As to conforming outwardly, and living your own life inwardly, I have not a very high opinion of that course.

If we were left solely to the wordy wit of legislators in Congress for our guidance, uncorrected by the seasonal experience and the effectual complaints of the people, America would not long retain her rank among the nations.

Every generation laughs at the old fashions, but follows religiously the new.

There is no odor so bad as that which arises from goodness tainted.

The Artist is he who detects and applies the law from observation of the works of Genius, whether of man or Nature. The Artisan is he who merely applies the rules which others have detected.

I am sorry to think that you do not get a man's most effective criticism until you provoke him. Severe truth is expressed with some bitterness.

Little is to be expected of that day, if it can be called a day, to which we are not awakened by our Genius, but by the mechanical nudgings of some servitor, are not awakened by our own newly acquired force and aspirations from within, accompanied by the undulations of celestial music, instead of factory bells, and a fragrance filling the air—to a higher life than we fell asleep from; and thus the darkness bear its fruit, and prove itself to be good, no less than the light. That man who does not believe that each day contains an earlier, more sacred, and auroral hour than he has yet profaned, has despaired of life, and is pursuing a descending and darkening way.

The mass of men lead lives of quiet desperation.

I do not know but thoughts written down thus in a journal might be printed in the same form with greater advantage than if the related ones were brought together into separate essays. They are now allied to life, and are seen by the reader not to be far-fetched. It is more simple, less artful. I feel that in the other case I should have no proper frame for my sketches. Mere facts and names and dates communicate more than we suspect....Perhaps I can never find so good a setting for my thoughts as I shall thus have taken them out of. The crystal never sparkles more brightly than in the cavern.

Is not disease the rule of existence? There is not a lily pad floating on the river but has been riddled by insects. Almost every shrub and tree has its gall, oftentimes esteemed its chief ornament and hardly to be distinguished from the fruit. If misery loves company, misery has company enough. Now, at midsummer, find me a perfect leaf or fruit.

It is an interesting question how far men would retain their relative rank if they were divested of their clothes.

The earth is not a mere fragment of dead history, stratum upon stratum like the leaves of a book, to be studied by geologists and antiquaries chiefly, but living poetry like the leaves of a tree, which precede flowers and fruit—not a fossil earth, but a living earth; compared with whose great central life all animal and vegetable life is merely parasitic. Its throes will heave our exuviae from their graves.

You must not blame me if I do talk to the clouds.

Whether the flower looks better in the nosegay than in the meadow where it grew and we had to wet our feet to get it! Is the scholastic air any advantage?

What does education often do? It makes a straight-cut ditch of a free, meandering brook.

How could youths better learn to live than by at once trying the experiment of living?

I should not talk so much about myself if there were anybody else whom I knew as well.

The light which puts out our eyes is darkness to us. Only that day dawns to which we are awake. There is more day to dawn. The sun is but a morning star.

I say, beware of all enterprises that require new clothes, and not rather a new wearer of clothes.

The broadest and most prevalent error requires the most disinterested virtue to sustain it.

Being is the great explainer.

It is easier to sail many thousand miles through cold and storm and cannibals, in a government ship, with five hundred men and boys to assist one, than it is to explore the private sea, the Atlantic and Pacific Ocean of one's being alone....It is not worth the while to go round the world to count the cats in Zanzibar.

My facts shall be falsehoods to the common sense. I would so state facts that they shall be significant, shall be myths or mythologic. Facts which the mind perceived, thoughts which the body thought—with these I deal.

The words which express our faith and piety are not definite; yet they are significant and fragrant like frankincense to superior natures.

By avarice and selfishness, and a groveling habit, from which none of us is free, of regarding the soil as property, or the means of acquiring property chiefly, the landscape is deformed, husbandry is degraded with us, and the farmer leads the meanest of lives. He knows Nature but as a robber.

Farmers are respectable and interesting to me in proportion as they are poor.

We worship not the Graces, nor the Parcae, but Fashion. She spins and weaves and cuts with full authority. The head monkey at Paris puts on a traveler's cap, and all the monkeys in America do the same.

The perch swallows the grub-worm, the pickerel swallows the perch, and the fisherman swallows the pickerel; and so all the chinks in the scale of being are filled.

One of the most attractive things about the flowers is their beautiful reserve.

I have found it to be the most serious objection to coarse labors long continued, that they compelled me to eat and drink coarsely also.

The American has dwindled into an Odd Fellow—one who may be known by the development of his organ of gregariousness.

To say that a man is your Friend, means commonly no more than this, that he is not your enemy. Most contemplate only what would be the accidental and trifling advantages of Friendship, as that the Friend can assist in time of need by his substance, or his influence, or his counsel....Even the utmost goodwill and harmony and practical kindness are not sufficient for Friendship, for Friends do not live in harmony merely, as some say, but in melody.

I suppose you think that persons who are as old as your father and myself are always thinking about very grave things, but I know that we are meditating the same old themes that we did when we were ten years old, only we go more gravely about it.

It seems to me that the god that is commonly worshipped in civilized countries is not at all divine, though he bears a divine name, but is the overwhelming authority and respectability of mankind combined. Men reverence one another, not yet God.

If I knew for a certainty that a man was coming to my house with the conscious design of doing me good, I should run for my life.

Goodness is the only investment that never fails.

Government is at best but an expedient; but most governments are usually, and all governments are sometimes, inexpedient. The objections which have been brought against a standing army, and they are many and weighty, and deserve to prevail, may also at last be brought against a standing government.

This American government—what is it but a tradition, though a recent one, endeavoring to transmit itself unimpaired to posterity, but each instant losing some of its integrity? It has not the vitality and force of a single living man; for a single man can bend it to his will.

The government of the world I live in was not framed, like that of Britain, in after-dinner conversations over the wine.

When I hear the hypercritical quarreling about grammar and style, the position of the particles, etc., etc., stretching or contracting every speaker to certain rules of theirs…I see that they forget that the first requisite and rule is that expression shall be vital and natural, as much as the voice of a brute or an interjection: first of all, mother tongue; and last of all, artificial or father tongue. Essentially your truest poetic sentence is as free and lawless as a lamb's bleat.

Any fool can make a rule
And every fool will mind it.

He who distinguishes the true savor of his food can never be a glutton; he who does not cannot be otherwise.

What right have I to grieve, who have not ceased to wonder?

We are made happy when reason can discover no occasion for it. The memory of some past moments is more persuasive than the experience of present ones. There have been visions of such breadth and brightness that these motes were invisible in their light.

Measure your health by your sympathy with morning and spring. If there is no response in you to the awakening of nature—if the prospect of an early morning walk does not banish sleep, if the warble of the first bluebird does not thrill you—know that the morning and spring of your life are past. Thus may you feel your pulse.

I went to the woods because I wished to live deliberately, to front only the essential facts of life, and see if I could not learn what it had to teach, and not, when I came to die, discover that I had not lived.…I wanted to live deep and suck out all the marrow of life, to live so sturdily and Spartan-like as to put to rout all that was not life, to cut a broad swath and shave close, to drive life into a corner, and reduce it to its lowest terms.

I had three chairs in my house; one for solitude, two for friendship, three for society.

Should not every apartment in which man dwells be lofty enough to create some obscurity overhead, where flickering shadows may play at evening about the rafters?

Nowadays the host does not admit you to *his* hearth, but has got the mason to build one for yourself somewhere in his alley, and hospitality is the art of *keeping* you at the greatest distance.

If you have built castles in the air, your work need not be lost; that is where they should be. Now put the foundations under them.

The man who does not betake himself at once and desperately to sawing is called a loafer, though he may be knocking at the doors of heaven all the while.

The man who goes alone can start today; but he who travels with another must wait till that other is ready, and it may be a long time before they get off.

Write while the heat is in you....The writer who postpones the recording of his thoughts uses an iron which has cooled to burn a hole with. He cannot inflame the minds of his audience.

What is peculiar in the life of a man consists not in his obedience, but his opposition, to his instincts. In one direction or another he strives to live a supernatural life.

The way in which men cling to old institutions after the life has departed out of them, and out of themselves, reminds me of those monkeys which cling by their tails—aye, whose tails contract about the limbs, even the dead limbs, of the forest, and they hang suspended beyond the hunter's reach long after they are dead. It is of no use to argue with such men. They have not an apprehensive intellect, but merely, as it were a prehensile tail.

Wherever a man goes, men will pursue him and paw him with their dirty institutions, and, if they can, constrain him to belong to their desperate odd-fellow society.

The laboring man has not leisure for a true integrity day by day.

The knowledge of an unlearned man is living and luxuriant like a forest, but covered with mosses and lichens and for the most part inaccessible and going to waste; the knowledge of the man of science is like timber collected in yards for public works, which still supports a green sprout here and there, but even this is liable to dry rot.

A man is rich in proportion to the number of things which he can afford to let alone.

A lake is the landscape's most beautiful and expressive feature. It is earth's eye; looking into which the beholder measures the depth of his own nature.

The lawyer's truth is not Truth, but consistency or a consistent expediency.

He enjoys true leisure who has time to improve his soul's estate.

A broad margin of leisure is as beautiful in a man's life as in a book. Haste makes waste, no less in life than in housekeeping. Keep the time, observe the hours of the universe, not of the cars. What are threescore years and ten hurriedly and coarsely lived to moments of divine leisure in which your life is coincident with the life of the universe?

Most of the luxuries and many of the so-called comforts of life are not only not indispensable, but positive hindrances to the elevation of mankind.

The mass never comes up to the standard of its best member, but on the contrary degrades itself to a level with the lowest.

In the midst of this chopping sea of civilized life, such are the clouds and storms and quicksands and thousand-and-one items to be allowed for, that a man has to live, if he would not founder and go to the bottom and not make his port at all, by dead reckoning, and he must be a great calculator indeed who succeeds.

We are eager to tunnel under the Atlantic and bring the Old World some weeks nearer to the New; but perchance the first news that will leak through into the broad, flapping American ear will be that the Princess Adelaide has the whooping cough.

The youth gets together his materials to build a bridge to the moon, or, perchance, a palace or temple on the earth, and, at length, the middle-aged man concludes to build a woodshed with them.

A minority is powerless while it conforms to the majority; it is not even a minority then; but it is irresistible when it clogs by its whole weight.

Our whole life is startlingly moral. There is never an instant's truce between virtue and vice.

A name pronounced is the recognition of the individual to whom it belongs. He who can pronounce my name aright, he can call me, and is entitled to my love and service.

If the fairest features of the landscape are to be named after men, let them be the noblest and worthiest men alone.

Nations! What are nations? Tartars! and Huns! and Chinamen! Like insects they swarm. The historian strives in vain to make them memorable. It is for want of a man that there are so many men. It is individuals that populate the world.

We can never have enough of nature. We must be refreshed by the sight of inexhaustible vigor, vast and titanic features, the sea-coast with its wrecks, the wilderness with its living and its decaying trees, the thundercloud, and the rain which lasts three weeks and produces freshets. We need to witness our own limits transgressed, and some life pasturing freely where we never wander.

To a philosopher all *news*, as it is called, is gossip, and they who edit it and read it are old women over their tea.

I have no time to read newspapers. If you chance to live and move and have your being in that thin stratum in which the events which make the news transpire—thinner than the paper on which it is printed—then these things will fill the world for you; but if you soar above or dive below that plane, you cannot remember nor be reminded of them.

You know about a person who deeply interests you more than you can be told. A look, a gesture, an act, which to everybody else is insignificant tells you more about that one than words can.

Yet some can be patriotic who have no *self-respect*, and sacrifice the greater to the less. They love the soil which makes their graves, but have no sympathy with the spirit which may still animate their clay. Patriotism is a maggot in their heads.

To be a philosopher is not merely to have subtle thoughts, nor even to found a school, but so to love wisdom as to live according to its dictates a life of simplicity, independence, magnanimity, and trust. It is to solve some of the problems of life, not only theoretically, but practically.

A stereotyped but unconscious despair is concealed even under what are called the games and amusements of mankind.

The poet is a man who lives at last by watching his moods. An old poet comes at last to watch his moods as narrowly as a cat does a mouse.

A farmer, a hunter, a soldier, a reporter, even a philosopher, may be daunted; but nothing can deter a poet, for he is actuated by pure love. Who can predict his comings and goings? His business calls him out at all hours, even when doctors sleep.

We are not what we are, nor do we treat or esteem each other for such, but for what we are capable of being.

Give me the poverty that enjoys true wealth.

Under a government which imprisons any unjustly, the true place for a just man is also a prison.

I respect not his labors, his farm where everything has its price, who would carry the landscape, who would carry his God, to market, if he could get anything for him; who goes to market *for* his god as it is; on whose farm nothing grows free, whose fields bear no crops, whose meadows no flowers, whose trees no fruits, but dollars.

I quietly declare war with the State, after my fashion, though I will still make use and get advantage of her as I can, as is usual in such cases.

The purity men love is like the mists which envelope the earth, and not like the azure ether beyond.

To watch this crystal globe just sent from heaven to associate with me. While these clouds and this sombre drizzling weather shut all in, we two draw nearer and know one another. The gathering in of the clouds with the last rush and dying breath of the wind, and then the regular dripping of twigs and leaves the country o'er, the impression of inward comfort and sociableness, the drenched stubble and trees that drop beads on you as you pass, their dim outline seen through the rain on all sides drooping in sympathy with yourself. These are my undisputed territory. This is Nature's English comfort.

Read the best books first, or you may not have a chance to read them at all.

To read well, that is, to read true books in a true spirit, is a noble exercise, and one that will task the reader more than any other exercise which the customs of the day esteem. It requires a training such as the athletes underwent, the steady intention almost of the whole life to this object.

The greater part of what my neighbors call good I believe in my soul to be bad, and if I repent of anything, it is very likely to be my good behavior. What demon possessed me that I behaved so well? You may say the wisest thing you can, old man,—you who have lived seventy years, not without honor of a kind,—I hear an irresistible voice which invites me away from all that.

I believe that what so saddens the reformer is not his sympathy with his fellows in distress, but, though he be the holiest son of God, is his private ail. Let this be righted, let the spring come to him, the morning rise over his couch, and he will forsake his generous companions without apology.

Make the most of your regrets; never smother your sorrow, but tend and cherish it till it come to have a separate and integral interest. To regret deeply is to live afresh.

What is called resignation is confirmed desperation.

The rich man…is always sold to the institution which makes him rich. Absolutely speaking, the more money, the less virtue.

I was born upon thy bank, river,
My blood flows in thy stream,
And thou meanderest forever
At the bottom of my dream.

The success of great scholars and thinkers is commonly a courtier-like success, not kingly, not manly.

If we knew all the laws of Nature, we should need only one fact, or the description of one actual phenomenon, to infer all the particular results at that point. Now we know only a few laws, and our result is vitiated, not, of course, by any confusion or irregularity in Nature, but by our ignorance of essential elements in the calculation. Our notions of law and harmony are commonly confined to those instances which we detect; but the harmony which results from a far greater number of seemingly conflicting, but really concurring, laws, which we have not detected, is still more wonderful. The particular laws are as our points of view, as, to the traveler, a mountain outline varies with every step, and it has an infinite number of profiles, though absolutely but one form. Even when cleft or bored through it is not comprehended in its entireness.

Live in each season as it passes; breathe the air, drink the drink, taste the fruit, and resign yourself to the influences of each. Let them be your only diet drink and botanical medicines.

Public opinion is a weak tyrant compared with our own private opinion. What a man thinks of himself, that it is which determines, or rather indicates, his fate.

To affect the quality of the day, that is the highest of arts. Every man is tasked to make his life, even in its details, worthy of the contemplation of his most elevated and critical hour.

Nay, be a Columbus to whole new continents and worlds within you, opening new channels, not of trade, but of thought. Every man is the lord of a realm beside which the earthly empire of the Czar is but a petty state, a hummock left by the ice.

The finest qualities of our nature, like the bloom on fruits, can be preserved only by the most delicate handling. Yet we do not treat ourselves nor one another thus tenderly.

I have been breaking silence these twenty-three years and have hardly made a rent in it.

Silence is the universal refuge, the sequel to all dull discourses and all foolish acts, a balm to our every chagrin, as welcome after satiety as after disappointment; that background which the painter may not daub, be he master or bungler, and which, however awkward a figure we may have made in the foreground, remains ever our inviolable asylum, where no indignity can assail, no personality can disturb us.

Simplicity, simplicity, simplicity! I say, let your affairs be as two or three, and not a hundred or a thousand; instead of a million count half a dozen, and keep your accounts on your thumb-nail.

We cannot well do without our sins; they are the highway of our virtue.

After the first blush of sin comes its indifference.

I only desire sincere relations with the worthiest of my acquaintance, that they may give me an opportunity once in a year to speak the truth.

Talk about slavery! It is not the peculiar institution of the South. It exists wherever men are bought and sold, wherever a man allows himself to be made a mere thing or a tool, and surrenders his inalienable rights of reason and conscience. Indeed, this slavery is more complete than that which enslaves the body alone....I never yet met with, or heard of, a judge who was not a slave of this kind, and so the finest and most unfailing weapon of injustice. He fetches a slightly higher price than the black men only because he is a more valuable slave.

What men call social virtues, good fellowship, is commonly but the virtue of pigs in a litter, which lie close together to keep each other warm.

I never found the companion that was so companionable as solitude.

It would give me such joy to know that a friend had come to see me, and yet that pleasure I seldom if ever experience.

However intense my experience, I am conscious of the presence and criticism of a part of me, which, as it were, is not a part of me, but a spectator, sharing no experience, but taking note of it, and that is no more I than it is you. When the play, it may be the tragedy, of life is over, the spectator goes his way. It was a kind of fiction, a work of the imagination only, so far as he was concerned.

Speech is for the convenience of those who are hard of hearing; but there are many fine things which we cannot say if we have to shout.

The stars are the apexes of what triangles!

I just looked up at a fine twinkling star and thought that a voyager whom I know, now many days' sail from this coast, might possibly be looking up at that same star with me.

There will never be a really free and enlightened State until the State comes to recognize the individual as a higher and independent power, from which all its own power and authority are derived, and treats him accordingly. I please myself with imagining a State at last which can afford to be just to all men, and to treat the individual with respect as a neighbor; which even would not think it inconsistent with its own repose if a few went to live aloof from it, not meddling with it, nor embraced by it, who fulfilled all the duties of neighbors and fellow-men. A State which bore this kind of fruit, and suffered it to drop off as fast as it ripened, would prepare the way for a still more perfect and glorious State, which also I have imagined, but not yet anywhere seen.

I was never molested by any person but those who represented the State.

If the day and the night are such that you greet them with joy, and life emits a fragrance like flowers and sweet-scented herbs, is more elastic, more starry, more immortal—that is your success. All nature is your congratulation, and you have cause momentarily to bless yourself.

The sun is but a morning star.

We are paid for our suspicions by finding what we suspected.

If a thousand men were not to pay their tax-bills this year, that would not be a violent and bloody measure, as it would be to pay them, and enable the State to commit violence and shed innocent blood. This is, in fact, the definition of a peaceable revolution, if any such is possible.

Our inventions are wont to be pretty toys, which distract our attention from serious things. They are but improved means to an unimproved end.

If I were confined to a corner of a garret all my days, like a spider, the world would be just as large to me while I had my thoughts about me.

Having each some shingles of thought well dried, we sat and whittled them.

As if you could kill time without injuring eternity.

But lo! men have become the tools of their tools.

That devilish Iron Horse, whose ear-rending neigh is heard through-out the town, has muddied the Boiling Spring with his foot, and he it is that has browsed off all the woods on Walden shore, that Trojan horse, with a thousand men in his belly, introduced by mercenary Greeks! Where is the country's champion, the Moore of Moore Hall, to meet him at the Deep Cut and thrust an avenging lance between the ribs of the bloated pest?

He who is only a traveler learns things at second-hand and by the halves, and is poor authority. We are most interested when science reports what those men already know practically or instinctively, for that alone is a true *humanity*, or account of human experience.

It takes two to speak the truth—one to speak and another to hear.

They who know of no purer sources of truth, who have traced up its stream no higher, stand, and wisely stand, by the Bible and the Constitution, and drink at it there with reverence and humility; but they who behold where it comes trickling into this lake or that pool, gird up their loins once more, and continue their pilgrimage toward its fountainhead.

We shall see but little way if we require to understand what we see. How few things can a man measure with the tape of his understanding! How many greater things might he be seeing in the meanwhile!

The universe is wider than our views of it.

At the same time that we are earnest to explore and learn all things, we require that all things be mysterious and unexplorable, that land and sea be infinitely wild, unsurveyed and unfathomed by us because unfathomable.

I have no doubt that it is a part of the destiny of the human race, in its gradual improvement, to leave off eating animals, as surely as the savage tribes have left off eating each other when they came in contact with the more civilized.

Every man who has ever been earnest to preserve his higher or poetic faculties in the best condition has been particularly inclined.

One farmer says to me, "You cannot live on vegetable food solely, for it furnishes nothing to make bones with"; and so he religiously devotes a part of his day to supplying his system with the raw material of bones; walking all the while he talks behind his oxen, which, with vegetable-made bones, jerk him and his lumbering plow along in spite of every obstacle.

There are nine hundred and ninety-nine patrons of virtue to one virtuous man.

All voting is a sort of gaming, like checkers or backgammon, with a slight moral tinge to it, a playing with right and wrong.

I have a deep sympathy with war, it so apes the gait and bearing of the soul.

Superfluous wealth can buy superfluities only. Money is not required to buy one necessary of the soul.

God is only the president of the day, and Webster is his orator.

We need the tonic of wildness, to wade sometimes in marshes where the bittern and the meadow-hen lurk, and hear the booming of the snipe; to smell the whispering sedge where only some wilder and more solitary fowl builds her nest, and the mink crawls with its belly close to the ground.

In wildness is the preservation of the world.

Many of the phenomena of Winter are suggestive of an inexpressible tenderness and fragile delicacy. We are accustomed to hear this king described as a rude and boisterous tyrant; but with the gentleness of a lover he adorns the tresses of Summer.

Sometimes we are inclined to class those who are once-and-a-half witted with the half-witted, because we appreciate only a third part of their wit.

It requires nothing less than a chivalric feeling to sustain a conversation with a lady.

The volatile truth of our words should continually betray the inadequacy of the residual statement.

The really efficient laborer will be found not to crowd his day with work, but will saunter to his task surrounded by a wide halo of ease and leisure.

The cost of a thing is the amount of what I will call life which is required to be exchanged for it, immediately or in the long run.

A perfectly healthy sentence, it is true, is extremely rare. For the most part we miss the hue and fragrance of the thought; as if we could be satisfied with the dews of the morning or evening without their colors, or the heavens without their azure.

Alexis de Tocqueville

(1805-59)
French social philosopher

Nothing is quite so wretchedly corrupt as an aristocracy which has lost its power but kept its wealth and which still has endless leisure to devote to nothing but banal enjoyments. All its great thoughts and passionate energy are things of the past, and nothing but a host of petty, gnawing vices now cling to it like worms to a corpse.

Among the laws controlling human societies there is one more precise and clearer, it seems to me, than all the others. If men are to remain civilized or to become civilized, the art of association must develop and improve among them at the same speed as equality of conditions spreads.

In democratic ages men rarely sacrifice themselves for another, but they show a general compassion for all the human race. One never sees them inflict pointless suffering, and they are glad to relieve the sorrows of others when they can do so without much trouble to themselves. They are not disinterested, but they are gentle.

I cannot help fearing that men may reach a point where they look on every new theory as a danger, every innovation as a toilsome trouble, every social advance as a first step toward revolution, and that they may absolutely refuse to move at all for fear of being carried off their feet. The prospect really does frighten me that they may finally become so engrossed in a cowardly love of immediate pleasures that their interest in their own future and in that of their descendants may vanish, and that they will prefer tamely to follow the course of their destiny rather than make a sudden energetic effort necessary to set things right.

Despotism may govern without faith, but liberty cannot....How is it possible that society should escape destruction if the moral tie is not strengthened in proportion as the political tie is relaxed? And what can be done with a people who are their own masters if they are not submissive to the Deity?

I do not find fault with equality for drawing men into the pursuit of forbidden pleasures, but for absorbing them entirely in the search for the pleasures that are permitted.

It is the dissimilarities and inequalities among men which give rise to the notion of honor; as such differences become less, it grows feeble; and when they disappear, it will vanish too.

However energetically society in general may strive to make all the citizens equal and alike, the personal pride of each individual will always make him try to escape from the common level, and he will form some inequality somewhere to his own profit.

Not only does democracy make every man forget his ancestors, but also clouds their view of their descendants and isolates them from their contemporaries. Each man is for ever thrown back on himself alone, and there is danger that he may be shut up in the solitude of his own heart.

The genius of democracies is seen not only in the great number of new words introduced but even more in the new ideas they express.

The best laws cannot make a constitution work in spite of morals; morals can turn the worst laws to advantage. That is a commonplace truth, but one to which my studies are always bringing me back. It is the central point in my conception. I see it at the end of all my reflections.

By and large the literature of a democracy will never exhibit the order, regularity, skill, and art characteristic of aristocratic literature; formal qualities will be neglected or actually despised. The style will often be strange, incorrect, overburdened, and loose, and almost always strong and bold. Writers will be more anxious to work quickly than to perfect details. Short works will be commoner than long books, wit than erudition, imagination than depth. There will be a rude and untutored vigor of thought with great variety and singular fecundity. Authors will strive to astonish more than to please, and to stir passions rather than to charm taste.

Though it is very important for man as an individual that his religion should be true, that is not the case for society. Society has nothing to fear or hope from another life; what is most important for it is not that all citizens profess the true religion but that they should profess religion.

When an opinion has taken root in a democracy and established itself in the minds of the majority, if afterward persists by itself, needing no effort to maintain it since no one attacks it. Those who at first rejected it as false come in the end to adopt it as accepted, and even those who still at the bottom of their hearts oppose it keep their views to themselves, taking great care to avoid a dangerous and futile contest.

The main business of religions is to purify, control, and restrain that excessive and exclusive taste for well-being which men acquire in times of equality.

Grant me thirty years of equal division of inheritances and a free press, and I will provide you with a republic.

It is almost never when a state of things is the most detestable that it is smashed, but when, beginning to improve, it permits men to breathe, to reflect, to communicate their thoughts with each other, and to gauge by what they already have the extent of their rights and their grievances. The weight, although less heavy, seems then all the more unbearable.

What is most important for democracy is not that great fortunes should not exist, but that great fortunes should not remain in the same hands. In that way there are rich men, but they do not form a class.

Consider any individual at any period of his life, and you will always find him preoccupied with fresh plans to increase his comfort. Do not talk to him about the interests and rights of the human race; that little private business of his for the moment absorbs all his thoughts, and he hopes that public disturbances can be put off to some other time.

Trade is the natural enemy of all violent passions. Trade loves moderation, delights in compromise, and is most careful to avoid anger. It is patient, supple, and insinuating, only resorting to extreme measures in cases of absolute necessity. Trade makes men independent of one another and gives them a high idea of their personal importance: it leads them to want to manage their own affairs and teaches them to succeed therein. Hence it makes them inclined to liberty but disinclined to revolution.

There are two things which will always be very difficult for a democratic nation: to start a war and to end it.

Leo Tolstoy

(1828-1910)
Russian novelist, philosopher

The chief difference between words and deeds is that words are always intended for men for their approbation, but deeds can be done only for God.

To say that a work of art is good, but incomprehensible to the majority of men, is the same as saying of some kind of food that it is very good but that most people can't eat it.

The Brahmins say that in their books there are many predictions of times in which it will rain. But press those books as strongly as you can, you can not get out of them a drop of water. So you can not get out of all the books that contain the best precepts the smallest good deed.

The changes in our life must come from the impossibility to live otherwise than according to the demands of our conscience…not from our mental resolution to try a new form of life.

A Frenchman is self-assured because he regards himself personally both in mind and body as irresistibly attractive to men and women. An Englishman is self-assured as being a citizen of the best-organized state in the world and therefore, as an Englishman, always knows what he should do and knows that all he does as an Englishman is undoubtedly correct. An Italian is self-assured because he is excitable and easily forgets himself and other people. A Russian is self-assured just because he knows nothing and does not want to know anything, since he does not believe that anything can be known. The German's self-assurance is worst of all, stronger and more repulsive than any other, because he imagines that he knows the truth—science—which he himself has invented but which is for him the absolute truth.

All happy families resemble one another, but each unhappy family is unhappy in its own way.

The best generals I have known were...stupid or absent-minded men....Not only does a good army commander not need any special qualities, on the contrary he needs the absence of the highest and best human attributes—love, poetry, tenderness, and philosophic inquiring doubt. He should be limited, firmly convinced that what he is doing is very important (otherwise he will not have sufficient patience), and only then will he be a brave leader. God forbid that he should be humane, should love, or pity, or think of what is just and unjust.

Though it is possible to utter words only with the intention to fulfill the will of God, it is very difficult not to think about the impression which they will produce on men and not to form them accordingly. But deeds you can do quite unknown to men, only for God. And such deeds are the greatest joy that a man can experience.

In quiet and untroubled times it seems to every administrator that it is only by his efforts that the whole population under his rule is kept going, and in this consciousness of being indispensable every administrator finds the chief reward of his labor and efforts. While the sea of history remains calm the ruler-administrator in his frail bark, holding on with a boat hook to the ship of the people and himself moving, naturally imagines that his efforts move the ship he is holding on to. But as soon as a storm arises and the sea begins to heave and the ship to move, such a delusion is no longer possible. The ship moves independently with its own enormous motion, the boat hook no longer reaches the moving vessel, and suddenly the administrator, instead of appearing a ruler and a source of power, becomes an insignificant, useless, feeble man.

In historic events, the so-called great men are labels giving names to events, and like labels they have but the smallest connection with the event itself. Every act of theirs, which appears to them an act of their own will, is in an historical sense involuntary and is related to the whole course of history and predestined from eternity.

Man lives consciously for himself, but is an unconscious instrument in the attainment of the historic, universal, aims of humanity.

Hypocrisy in anything whatever may deceive the cleverest and most penetrating man, but the least wide-awake of children recognizes it, and is revolted by it, however ingeniously it may be disguised.

I sit on a man's back, choking him and making him carry me, and yet assure myself and others that I am very sorry for him and wish to ease his lot by all possible means—except by getting off his back.

Love is life. All, everything that I understand, I understand only because I love. Everything is, everything exists, only because I love. Everything is united by it alone. Love is God, and to die means that I, a particle of love, shall return to the general and eternal source.

True science investigates and brings to human perception such truths and such knowledge as the people of a given time and society consider most important. Art transmits these truths from the region of perception to the region of emotion.

Miguel de Unamuno

(1864-1936)
Spanish philosophical writer

When a thing is said to be not worth refuting you may be sure that either it is flagrantly stupid—in which case all comment is superfluous—or it is something formidable, the very crux of the problem.

The only way to give finality to the world is to give it consciousness.

Man dies of cold, not of darkness.

We need God, not in order to understand the *why*, but in order to feel and sustain the ultimate *wherefore*, to give a meaning to the universe.

To fall into a habit is to begin to cease to be.

For it is the suffering flesh, it is suffering, it is death, that lovers perpetuate upon the earth. Love is at once the brother, son, and father of death, which is its sister, mother, and daughter. And thus it is that in the depth of love there is a depth of eternal despair, out of which springs hope and consolation.

To love with the spirit is to pity, and he who pities most loves most.

Science is a cemetery of dead ideas.

The skeptic does not mean him who doubts, but him who investigates or researches, as opposed to him who asserts and thinks that he has found.

There is no true love save in suffering, and in this world we have to choose either love, which is suffering, or happiness....Man is the more man—that is, the more divine—the greater his capacity for suffering, or rather, for anguish.

Cure yourself of the affliction of caring how you appear to others. Concern yourself only with how you appear before God, concern yourself only with the idea that God may have of you.

Giambattista Vico

(1688-1744)
Italian philosopher, historian

Common sense is judgment without reflection, shared by an entire class, an entire nation, or the entire human race.

Uniform ideas originating among entire peoples unknown to each other must have a common ground of truth.

Men first feel necessity, then look for utility, next attend to comfort, still later amuse themselves with pleasure, thence grow dissolute in luxury, and finally go mad and waste their substance.

The nature of peoples is first crude, then severe, then benign, then delicate, finally dissolute.

The universal principle of etymology in all languages: words are carried over from bodies and from the properties of bodies to express the things of the mind and spirit. The order of ideas must follow the order of things.

Metaphysics abstracts the mind from the senses, and the poetic faculty must submerge the whole mind in the senses. Metaphysics soars up to universals, and the poetic faculty must plunge deep into particulars.

It is true that men themselves made this world of nations...but this world without doubt has issued from a mind often diverse, at times quite contrary, and always superior to the particular ends that men had proposed to themselves.

But the nature of our civilized minds is so detached from the senses, even in the vulgar, by abstractions corresponding to all the abstract terms our languages abound in, and so refined by the art of writing, and as it were spiritualized by the use of numbers, because even the vulgar know how to count and reckon, that it is naturally beyond our power to form the vast image of this mistress called "Sympathetic Nature."

Voltaire

(1694-1778)
French philosopher, author

If we do not find anything very pleasant, at least we shall find something new.

It is not known precisely where angels dwell—whether in the air, the void, or the planets. It has not been God's pleasure that we should be informed of their abode.

To the living we owe respect, but to the dead we owe only the truth.

The consolation of deaf people is to read, and sometimes to scribble.

Men who are occupied in the restoration of health to other men, by the joint exertion of skill and humanity, are above all the great of the earth. They even partake of divinity, since to preserve and renew is almost as noble as to create.

I have only ever made one prayer to God, a very short one: "O Lord, make my enemies ridiculous." And God granted it.

The best is the enemy of the good.

I disapprove of what you say, but I will defend to the death your right to say it.

God is not on the side of the big battalions, but on the side of those who shoot best.

Governments need to have both shepherds and butchers.

Men hate the individual whom they call avaricious only because nothing can be gained from him.

History should be written as philosophy.

If there were only one religion in England there would be danger of despotism, if there were two, they would cut each other's throats, but there are thirty, and they live in peace and happiness.

It is not love that should be depicted as blind, but self-love.

Anyone who seeks to destroy the passions instead of controlling them is trying to play the angel.

One merit of poetry few persons will deny: it says more and in fewer words than prose.

I advise you to go on living solely to enrage those who are paying your annuities. It is the only pleasure I have left.

We must cultivate our own garden....When man was put in the garden of Eden he was put there so that he should work, which proves that man was not born to rest.

Shakespeare, who was considered the English Corneille, flourished at about the time of Lope de Vega. He had a strong and fertile genius, full of naturalness and sublimity, without the slightest spark of good taste or the least knowledge of the rules....After two hundred years most of the outlandish and monstrous ideas of this author have acquired the right to be considered sublime, and almost all modern authors have copied him....It does not occur to people that they should not copy him, and the lack of success of their copies simply makes people think that he is inimitable.

The husband who decides to surprise his wife is often very much surprised himself.

In general, the art of government consists in taking as much money as possible from one party of the citizens to give to the other.

They use thought only to justify their injustices, and speech only to disguise their thoughts.

Woe to the makers of literal translations, who by rendering every word weaken the meaning! It is indeed by so doing that we can say the letter kills and the spirit gives life.

Time, which alone makes the reputation of men, ends by making their defects respectable.

Alan Watts

(1915-73)
British-born U.S. philosopher, author

Zen…does not confuse spirituality with thinking about God while one is peeling potatoes. Zen spirituality is just to peel the potatoes.

Otto Weininger

(1880-1903)
Austrian philosopher

No men who really think deeply about women retain a high opinion of them; men either despise women or they have never thought seriously about them.

Alfred North Whitehead

(1861-1947)
British philosopher

Art is the imposing of a pattern on experience, and our aesthetic enjoyment is recognition of the pattern.

Life is an offensive, directed against the repetitious mechanism of the Universe.

Every philosophy is tinged with the colouring of some secret imaginative background, which never emerges explicitly into its train of reasoning.

There are no whole truths; all truths are half-truths. It is trying to treat them as whole truths that plays the devil.

William of Occam

(c.1285-c.1349)
English monk, philosopher

It is vain to do with more what can be done with less.

Ludwig Wittgenstein

(1889-1951)
Austrian philosopher

It is a dogma of the Roman Church that the existence of God can be proved by natural reason. Now this dogma would make it impossible for me to be a Roman Catholic. If I thought of God as another being like myself, outside myself, only infinitely more powerful, then I would regard it as my duty to defy him.

If you do know that *here is one hand*, we'll grant you all the rest.

A confession has to be part of your new life.

A philosopher who is not taking part in discussions is like a boxer who never goes into the ring.

Death is not an event in life: we do not live to experience death. If we take eternity to mean not infinite temporal duration but timelessness, then eternal life belongs to those who live in the present.

The face is the soul of the body.

Like everything metaphysical the harmony between thought and reality is to be found in the grammar of the language.

What has history to do with me? Mine is the first and only world! I want to report how I find the world. What others have told me about the world is a very small and incidental part of my experience. I have to judge the world, to measure things.

I am my world.

No one likes having offended another person; hence everyone feels so much better if the other person doesn't show he's been offended. Nobody likes being confronted by a wounded spaniel. Remember that. It is much easier patiently—and tolerantly—to avoid the person you have injured than to approach him as a friend. You need courage for that.

You must always be puzzled by mental illness. The thing I would dread most, if I became mentally ill, would be your adopting a common sense attitude; that you could take it for granted that I was deluded.

Never stay up on the barren heights of cleverness, but come down into the green valleys of silliness.

Knowledge is in the end based on acknowledgment.

Language is a part of our organism and no less complicated than it.

Everyday language is a part of the human organism and is no less complicated than it.

Someone who knows too much finds it hard not to lie.

I sit astride life like a bad rider on a horse. I only owe it to the horse's good nature that I am not thrown off at this very moment.

Logic takes care of itself; all we have to do is to look and see how it does it.

The logic of the world is prior to all truth and falsehood.

Logic is prior to every experience—that something is so.

For a *large* class of cases—though not for all—in which we employ the word "meaning" it can be defined thus: the meaning of a word is its use in the language.

It is one of the chief skills of the philosopher not to occupy himself with questions which do not concern him.

Philosophy is like trying to open a safe with a combination lock: each little adjustment of the dials seems to achieve nothing, only when everything is in place does the door open.

The real discovery is the one which enables me to stop doing philosophy when I want to.—The one that gives philosophy peace, so that it is no longer tormented by questions which bring *itself* into question.

We *regard* the photograph, the picture on our wall, as the object itself (the man, landscape, and so on) depicted there. This need not have been so. We could easily imagine people who did not have this relation to such pictures. Who, for example, would be repelled by photographs, because a face without color and even perhaps a face in reduced proportions struck them as inhuman.

A man will be imprisoned in a room with a door that's unlocked and opens inwards; as long as it does not occur to him to pull rather than push.

Don't get involved in partial problems, but always take flight to where there is a free view over the whole *single* great problem, even if this view is still not a clear one.

It is so characteristic, that just when the mechanics of reproduction are so vastly improved, there are fewer and fewer people who know how the music should be played.

Our civilization is characterized by the word "progress." Progress is its form rather than making progress being one of its features. Typically it constructs. It is occupied with building an ever more complicated structure. And even clarity is sought only as a means to this end, not as an end in itself. For me on the contrary clarity, perspicuity are valuable in themselves.

For a truly religious man nothing is tragic.

Not every religion has to have St. Augustine's attitude to sex. Why even in our culture marriages are celebrated in a church, everyone present knows what is going to happen that night, but that doesn't prevent it being a religious ceremony.

Man has to awaken to wonder—and so perhaps do peoples. Science is a way of sending him to sleep again.

The philosophical I is not the human being, not the human body or the human soul with the psychological properties, but the metaphysical subject, the boundary (not a part) of the world.

The philosophical self is not the human being, not the human body, or the human soul with which psychology deals, but rather the metaphysical subject, the limit of the world—not a part of it.

Nothing is so difficult as not deceiving oneself.

Whereof one cannot speak, thereof one must be silent.

What can be said at all can be said clearly, and what we cannot talk about we must pass over in silence.

There are remarks that sow and remarks that reap.

Our greatest stupidities may be very wise.

If a person tells me he has been to the worst places I have no reason to judge him; but if he tells me it was his superior wisdom that enabled him to go there, then I know he is a fraud.

In order to be able to set a limit to thought, we should have to find both sides of the limit thinkable (i.e. we should have to be able to think what cannot be thought).

A man's thinking goes on within his consciousness in a seclusion in comparison with which any physical seclusion is an exhibition to public view.

You get tragedy where the tree, instead of bending, breaks.

When one is frightened of the truth…then it is never the *whole* truth that one has an inkling of.

Resting on your laurels is as dangerous as resting when you are walking in the snow. You doze off and die in your sleep.

It seems to me that, in every culture, I come across a chapter headed "Wisdom." And then I know exactly what is going to follow: "Vanity of vanities, all is vanity."

A new word is like a fresh seed sewn on the ground of the discussion.

Xenocrates

(396-315 B.C.)
Greek philosopher

I have often repented speaking, but never of holding my tongue.

Printed in the United States
711000003B